I0057051

WHEN
CREDENTIALS
CAUSE HARM

UNPACKING THE RISKS OF VERIFIABLE LEARNING AND WORK RECORDS

KELLY PAGE, PH.D.

LWYL Studio

Copyright © 2026 LWYL Studio

All rights reserved

No part of this book may be reproduced, or stored in a retrieval system, or transmitted in any form or by any means, electronic, mechanical, photocopying, recording, or otherwise, without express written permission of the publisher.

ISBN: 978-1-969861-00-0

Cover design by:

Anze Ban Virant, ABV Atelier Design

Interior Design by HMD Publishing

Printed in the United States of America

DEDICATION

To my codesigner in life, Russell,
Thank you for walking this adventure alongside me. Your unwavering support and inclusive spirit are woven into every page of this book. You inspire and support me daily to live what I/we love, shaping a world where our values and passions become our most remarkable co-creations.

To my greatest ally, Nicci,
Since as young as I could remember playing, learning, and navigating life in Australia, you've been by my side encouraging me. Thank you for your fierce support and sisterhood. You've shown me what it truly means to lift each other up, creating a world where women supporting women is how we thrive together.

With endless gratitude,
Kelly
X

ACKNOWLEDGMENT

This book was shaped by many critical conversations, generous provocations, and ongoing reflections with people who have long challenged the digital credentialing status quo, not just with critique, yet with care, courage, and clarity.

I am especially grateful to:

Dr. Sheryl Grant, for her unwavering commitment to questioning the narratives we inherit about learning, legitimacy, and recognition and for reminding us that not everything meaningful needs to be measured.

Dr. Krystal Rawls, whose insight into equity, race, and labor reframed how I think about systemic harm, and whose work continually calls us to center lived experience.

Dr. Brian Tinsley, whose care and commitment to ethical and justice-informed research of learning and work systems, and the people and organizations who lead and benefit from them.

Dr. Michael Torrence, for his leadership in bringing ethics and equity into real-world post-secondary education implementations of digital credentials, and for making space for hard truths.

Dr. Rupert Ward, for his critical lens on what counts as recognition, what's been lost in translation, and what frameworks for microcredentialing systems might still become.

Dr. Kim Hamilton Duffy, for her groundbreaking contributions to decentralized identity and privacy-preserving infrastructure. Her leadership has ensured that identity systems can empower rather than expose, grounded in privacy-first design.

Dr. Kerri Lemoie, for her foundational role in developing open, interoperable verifiable credential standards and open ecosystems to shape a future where trust is earned, not assumed.

Dr. Julie Keane, for modeling a community of practice ethos grounded in learning, inclusion, partnership, and deeply rooted questions about technology's role in public life.

Dr. Angela Constani, for her unwavering dedication to the development of skills and their validation through real-life, human experiences and her leadership in the field of bioscience.

Meena Naik, for her questioning of inequitable institutional systems, centering the challenges and exploring ways to create more equitable experiences, that are more than functional, they are just.

Brooke Liptz, for her deep thought partnership and her commitment to learner-centered innovation. Her work helped surface some of the most pressing tensions between platform design and human dignity.

Margot Griffith, for your depth of care and knowledge about skills, skills validation, and how to truly honor the learners experience as they develop and use their skills.

Wendy Palmer, for your commitment to building badging programs and experiences with care and story, which honor the individual and the organization.

Javier Motta-Mena, for his passion for all things about accessible and inclusive design, in learning and digital credential, and his care in what being accountable to each other means.

Kelly Rondu, for her care for all things worker-led, worker-centered, and worker-first, in post-secondary learning design, and how we create space to authentically learn from and with each.

Kymberly Lavigne-Hinkley, for your kindness, and leadership in building a world where students can both understand and express the value of their unique skills and achievements.

Darin Hobbs, for his deep commitment to creating pathways to opportunities, and changing the lives of individuals and their families through digital credentials.

Kate Giovacchini, for her leadership and expertise curating networks centering trust, learning, and a commitment to equitable design that is also accessible to all learners.

Serge Ravet, who has been at the forefront of open recognition and trust-based credentialing systems, long before they were trends and who continues to remind us that recognition is a social, not technical, act.

Don Presant, whose commitment to ethical, open, and people-centered credentialing tools has informed much of the thinking in these pages.

Multiple pilot and demonstration teams I've been fortunate to meet, learn with and care alongside, such as the ASU Trusted Learner Network, SOLO Network, Alabama Talent Triad, WGU Achievement Wallet, Accelerate Montana, ColoradoFWD, Community College System Behavioral Health, Workforce Boulder County, Pikes Peak Workforce

Center, Colorado DIrect Care Careers, ASU Student Worker Employment for Skills-Based Success, Central Ohio Talent Network, Pittsburgh Regional Upskilling Alliance's LER Initiative. For all their hard work designing, testing, trialing, demonstrating, piloting, and asking all the questions we need to, in order to make recognition and talent visibility through credential innovation in learning and work a human reality.

To the **Open Recognition Alliance**, whose work has seeded a global movement to reimagine recognition as a relationship with trust, care, and community at its heart.

Thank you all for the conversations, provocations, and solidarity. You helped give language to what many feel yet can't always name.

This book stands on the ground you live in every day. .

PREFACE

This book began with a quiet unease.

For years, we've heard the promises: Digital credentials and open badges will unlock opportunity. Learning and Employment Records (LERs)[1] will level the playing field. Verifiable Digital Credentials (VDCs) will restore trust, put power and data back in the hands of learners and workers. Together, LERs issued as VDCs will solve the talent gap.

We desperately want to believe these promises. Many of us work in the institutions, nonprofits, and startups building them. We sat in the working groups, and on the community of practice calls. We did the research and wrote the discovery papers. We gave talks about interoperability, *'meeting people where they are,'* and about the transformative potential of skills-rich records and credentialing systems.

Yet, alongside the enthusiasm, another story keeps unfolding, less visible, less marketable, yet deeply real.

We listened to immigrants and returning citizens who *are told* to prove *'employability'* without being allowed to narrate the complexity of their lives. To frontline workers and students who are told to download apps to claim 'opt-in' credentials that weren't really optional. To job seekers

---

1. We use the term Learning and Work Records (LWRs) throughout this book, instead of Learning and Employment Records (LERs). See the Introduction for an explanation. .

in need of human support being told to 'login' 'download' 'chat' or 'call' an AI-powered support line for information. To refugees and justice-impacted individuals who are told to enter intimate and highly sensitive details into a system they are not allowed to control. To learners and workers whose badges are of little value beyond the organizations and platforms that issued them, yet require much learner time to access, store, and manage them.

We saw systems built quickly, without any quality assurance testing, compliance audits, feedback, or transparency, because the investors and funders demanded scale. We saw reports and evaluations written by those with something to sell. We saw funders stepping over the boundary into design and ecosystem decisions without really understanding the work on the ground. We saw stories of harm, care, social labor, and cultural knowledge ignored because it wasn't easily credentialed, could delay scale or minimize the return on financial investment.

So we started asking questions.

▸ What if these tools, built to recognize and uplift, were also capable of harm?

▸ What if they reproduced and at scale, the very exclusions they claim to disrupt?

▸ What would it mean to name that harm, not to stop progress, yet to make it more honest, more accountable, and more human?

This book is an attempt to do just that. It is not a rejection of verifiable digital credentials or learning and work records.

It is a call to slow down, to be brave, and to ask: Who is really being recognized and credentialed? By whom? For what purpose? How is the system designed? Where does the data come from and go to? Who benefits? How is it governed? And most importantly, what are the real human consequences?

You will find no silver bullets in these pages or a road-map. You will find a collection of ideas and actions for doing better. For building with care. For codesigning systems **with** the people most impacted. For creating a digital verifiable record and credentialing infrastructure that reflects justice and social impact, not just efficiency, data collection, or to feed the data machine.

If you work in the digital credentialing, verifiable credentialing or the learning and work record ecosystem. Be it in policy, design, education, workforce development, philanthropy, or technology, this book is written for you. It is written not to assign blame, yet to invite responsibility. To transform good intentions into meaningful, shared governance and careful action. To ensure that digital verifiable records and credentials do not merely sort and rank people or waste their already scarce time, money, and attention, yet to really serve to recognize, restore, and to build respect.

Let this be our starting point:

Credentials carry power.
So let us build them with care.

Kelly Page.

CONTENTS

CHAPTER 06. Surveillance, Tracking, and Control 117

CHAPTER 07. Vendor Lock-In and Platform Dependence. .. .131

INTRODUCTION: THE PARADOX OF PROMISE AND PERIL

CHAPTER INSIGHTS

- ► Issuing learning and work records as verifiable digital credentials is rapidly expanding, creating both opportunities and risks
- ► These systems embody a fundamental tension between empowering and extracting value from people and communities
- ► Centering justice requires intentional language, and care-centered principles, and practices

In the last decade, the education and workforce sectors have seen a surge of interest in a new category of data systems: Learning and Work Records (LWRs) issued as Verifiable Digital Credentials (VDCs). These technologies promise to revolutionize how individuals represent their skills, learning and work experiences, and how we may verify them. They are often framed as a way to unlock economic mobility, improve system efficiency, and reduce bias in hiring, recruitment, and advancement activities.

Yet as we increasingly embed these technologies into public systems and social policy, it is vital to ask: *Who benefits, and who might be harmed?*

This book is a call for critical reflection and mindful action.

While LWRs issued as VDCs may seem like neutral tools, just data formats, just infrastructure, just code and protocols, they are deeply political. They carry the values, assumptions, and power dynamics of the people and institutions that design, fund, and deploy them. And when implemented without care, they can replicate and even amplify the inequities many are promising they can or may solve.

A TALE OF TWO FUTURES

Picture two scenarios for the same job seeker who completed Heating, Ventilation, and Air Conditioning (HVAC) training as part of a vocational education and reentry program for justice-impacted individuals (See Table 1).

Table 1. Job Seeking Scenarios

Scenario 1	Scenario 2
A job seeker shares with an employer a privacy-preserving digital verifiable credential that can be used to verify - Yes or No, a) yes, they hold a professional HVAC certificate, b) yes, they have validated skills needed for the job, and c) the year when the credential was issued.	A job seeker shares via link or QR code in an online job application, marketplace or via an employer verifier dashboard, their verifiable digital credential, allowing the system to capture all the data from their credential for employer review, without clear consent mechanisms or data controls.

The job seeker can choose to share or not the details of where they received the training, the type of program, and for how long the employer has access to their credential information.	The application or dashboard collects the name of the training program, the type as a reentry program, where they received the training, alongside all their skills data even if it is not relevant for the role being applied for.
A hiring manager reviews the verification report and moves them along in the hiring process.	A screening system which uses artificial intelligence to identify skills as well as at-risk candidates flags them as "high risk" based on this information and rejects their application before any human review or official background check.

© LWYL Studio (2025)

Both of these futures are technically possible with the same tools. The difference lies not just in the technology, yet in how it is governed, designed, and implemented.

This is the central paradox of LWRs issued as VDCs: while they promise empowerment through data self-sovereignty and privacy-preserving consent mechanisms, they simultaneously create new vectors for harm. These systems can become tools of surveillance, coercion, and exclusion. They can also enable ecosystem actors to compile comprehensive data profiles, aggregating more personally identifiable information, learning, and work data far beyond traditional privacy considerations, and at scale.

All of this can occur without informed consent from the individual learner and worker, just by scanning a QR code, sharing a link, or sharing a credential from their wallet unaware which of the above two systems they may be using and how they may be exposed to harm.

THE PROMISE: SKILLS-RICH RECOGNITION AND DATA SELF-SOVEREIGNTY

The idea of issuing a digital verifiable credential that a person can carry across jobs, sectors, and borders is powerful. In theory, it allows individuals to own and manage their data, prove their abilities, and gain access to learning and work opportunities they might otherwise be excluded from. Advocates see LWRs issued as VDCs as a way to:

▸ **Move beyond** resumes and degrees to a more granular, fair representation of what someone can do and the experiences they've had.

▸ **Reduce** time-to-hire, job-candidate matching, and onboarding delays in industries facing labor shortages.

▸ **Support** recognition of prior learning and experience, especially for more diverse and modern workforce models, learners and workers.

▸ **Empower** individuals by letting them choose what to share, with whom, and when.

These goals are laudable. They reflect real frustrations with outdated record and credentialing systems, which are often slow, opaque, and exclusionary.

Yet good intentions are not enough.

THE PERIL: CODIFYING HARM AND DEEPENING INEQUITY

Issuing learning and work records as verifiable digital credentials, if implemented without ethical, regulatory, and legal guardrails, can do real harm. They can:

▸ **Expose** people to surveillance by employers, states, and vendors.

- ▶ **Force** individuals to disclose sensitive or stigmatizing information and store them in permanent records on systems they may not have access to or knowledge of.

- ▶ **Entrench** biases by privileging certain record and credential issuers, decisions, formats, and data.

- ▶ **Create** new forms of digital and data redlining where people from marginalized communities lack access to the tools, skills, or ecosystems necessary to participate in informed ways, and to benefit from the systems.

Framing the issuance of LWRs as VDC as *empowering* often masks underlying power asymmetries. *Who gets to issue verifiable digital credentials? Who decides which ones are valid? Who governs the wallets and platforms and where any of our data is stored, for how long, and if it is permanent? And most importantly, how is it being used to feed systems of artificial intelligence without our informed consent.*

These questions are rarely asked, and even less frequently answered to those most impacted when the solutions don't live up to the promises being made.

THE NEED FOR A JUSTICE LENS

Learning and work records issued as verifiable digital credentials are too consequential to be evaluated solely on technical functionality or the business value. We must also apply a justice lens, a framework that prioritizes equity, fairness and the context expertise of people from marginalized communities. Why? Because the real lived social impact of these systems determine whether they expand or constrain human opportunity.

In this we must ask:

- **How** do these tools affect power relations between individual learners and workers and institutions such as state agencies, educational organizations, and/or employers?
- **What** assumptions are embedded in record and credential design, frameworks, and schemas?
- **How** are informed consent, issuance, revocation, and verification handled in practice?
- **Are** people from marginalized communities meaningfully involved in the governance, design, and research of these systems?
- **How are we** ensuring accountability and responsibility to the promises, the standards, and the human impact?

To do this, in this book we draw on a range of perspectives: technologists, ethicists, community organizers, workforce development practitioners, and people with lived experience. The very people with the context expertise navigating the job market and learning, and work systems under conditions of precarity.

WHAT THIS BOOK OFFERS

This book offers a critical and necessary examination of the promises and perils of verifiable digital record and credentialing systems, specifically focusing on the issuance of Learning and Work Records (LWRs) as Verifiable Digital Credentials (VDCs).

As government agencies, nonprofits, employers, and learning and work technology companies rush to modernize recognition and verification systems through digital infrastructure and tools, *When Credentials Cause Harm* steps back to ask: *What happens when these tools don't serve the people they claim to empower? What happens when they cause harm?*

Drawing from case studies, policy critique, system design analysis, demonstrations, pilots, and lived experience, the book explores how these technologies can inadvertently amplify inequity, misrepresent individuals, reinforce surveillance, enable mass data extraction, and create new forms of exclusion, particularly for learners and workers from marginalized communities.

This book also offers a way toward ethical, human-centered alternatives. It invites designers, funders, policymakers, and advocates to rethink what counts as learning and work, who gets to define value and participation, and how power in the form of learning and work documentation, skills-rich data and personally identifiable information (PII) flows through technology-driven record and credentialing systems.

With frameworks for equity-informed design, consentful data practices, and community-led governance grounded in codesign, *When Credentials Cause Harm* is both a cautionary tale and a hopeful guide for building tools that recognize, support and truly protect people.

A NOTE ON LANGUAGE AND PERSPECTIVE

Before we get started we want to draw your attention to the language and perspectives we have taken in this book.

Given the many different ways that groups of people directly impacted are named, as well as the many terms used to refer to innovations, we wanted to be clear with readers as to how we recognize people, especially learners and workers in the work, and what we mean we use the terms

Learning and Work Records (LWRs), and Verifiable Digital Credentials (VDCs).

Our language and perspectives are shown in Figure 1 and explained in the following sections. You can also find the definitions we use in the glossary in the appendix

We hope these offer an inclusive and expansive foundation as you read and discover the book's contents.

Figure 1. Language and Perspectives

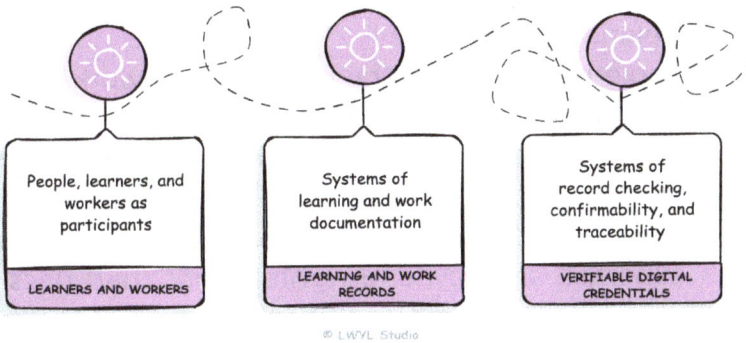

People, learners, and workers as participants	Systems of learning and work documentation	Systems of record checking, confirmability, and traceability
LEARNERS AND WORKERS	LEARNING AND WORK RECORDS	VERIFIABLE DIGITAL CREDENTIALS

© LWVL Studio

We recognize people, learners, and workers as participants

In this book we intentionally use the terms *'people,'* *'learners,'* and *'workers'* when referring to who these systems may serve or impact. When referring to people who participate in a project, program, a demonstration, or use a technology we refer to them as *'participants.'* In this we reject and avoid the use of the terms ~~'users'~~, ~~'customers'~~, ~~'consumers'~~, ~~'personas'~~, ~~'audiences'~~, or ~~'research subjects'~~.

▶ The terms *'people,'* *'learner,'* *'worker,'* and *'participant* reflect our commitment to acknowledging and centering the *people* who use, interact, manage, and contribute to technologies as *participants*. Learners and workers actively and passively participate in learning and work systems, contributing to the design of these systems through their choices, behavior, interactions, and the use of their data by organizations to inform the development of technology systems. For many pilots or demonstrations of new or emerging learning and work innovations, people also serve as active research participants, providing feedback, input, ideas, and solutions.

► The terms 'user', 'customer', 'consumer', 'persona', 'audience', and 'research subject', all center the systems, institutions, or researchers rather than the people themselves, obscuring the power imbalances inherent in these relationships and failing to actively acknowledge the people most affected by learning and work record and credentialing systems. It in particular further marginalized people from justice-impacted, immigrant, disabled, and economically marginalized communities, who deserve to be recognized as whole human beings with expertise, agency, and rights rather than passive recipients of technological solutions.

This choice reflects our centering of the power, rights and experiences of people whose stories and data is most frequently extracted and exploited, and often without agreement, informed consent, and least often protected. Our approach draws from critical race theory, feminist technology studies, and Indigenous data sovereignty frameworks. Equally important, it is grounded in the lived experiences and contextual expertise of people from marginalized communities whose voices too often go unheard or are exploited in technology-centered innovation development and system digital transformation.

Throughout this book we use the terms: people, learners, workers, and/or participants.

We refer to learning and work records as complex living systems of documentation

In this book we recognize *Learning and Work Records (LWRs)* as data rich records and record infrastructures that are complex living systems. They may be digitised, analogue, or experiential.

We intentionally use *'learning and work'* rather than ~~'learning and employment', 'education and employment' or 'learning and earning'~~ to recognize the full and diverse spectrum of valuable human contributions and labor, including care work, community organizing, creative pursuits, unpaid activities, and informal economic events that sustain lives and communities.

▸ The term *'learning,'* we use to acknowledge that people develop skills, knowledge, and expertise through community knowledge-sharing, peer mentorship, lived experience, apprenticeships, self-directed study, cultural transmission, family and friend interactions, and countless other experiences that extend far beyond traditional educational structures. The term ~~'education'~~ restricts our focus to formal institutional systems with credentialed instructors and structured curricula, while excluding the diverse ways people actually acquire knowledge and skills.

This recognition of learning is especially critical for people from justice-impacted, immigrant, disabled, and economically marginalized communities, whose knowledge systems and learning practices are often devalued or rendered invisible by formal educational institutions, yet represent essential forms of human development and skill-building that deserve recognition and protection.

- The term *'work,'* we use to acknowledge that people contribute value and build livelihoods through diverse experiences that extend far beyond traditional employment structures. This recognition is especially critical for people from justice-impacted, immigrant, disabled, and economically marginalized communities whose experiences inform this analysis. The term ~~'employment'~~ restricts the focus to formal job relationships with employers, while ~~'earning'~~ centers monetary compensation, both of which exclude many forms of work that are essential but often undervalued and/or unpaid.

- The term *'record'* in Learning and Work Records (LWRs) is used to mean any documentation, evidence, or trace of learning and work activities, not just formal institutional transcripts or employment histories. This may encompass official certificates and diplomas alongside community testimonials, project portfolios, peer endorsements, skill demonstrations, volunteer logs, documentation of care work, creative outputs, and any other artifact or asset that captures someone's knowledge, capabilities, or contributions. These records can be digital or physical, formal or informal, institutional or community-generated, standardized or uniquely personal. They are artifacts of one's life's activities, not merely paper, product, code or protocols to ship.

By embracing this expansive understanding of learning and work records, we acknowledge that people create and maintain many diverse forms of documentation about their learning and work throughout their lives, often outside of formal or digital systems, that people use to establish trust and communicate their abilities and experiences. We further recognize how we design, develop, and deploy these record systems, fundamentally shaping the lives and livelihoods of people, learners, and workers who live, navigate, learn, or work within them.

Throughout this book we use the terms: Learning and Work Records, LWRs or records of learning and work.

We refer to verifiable digital credentials as systems of record or credential checking, confirmability, and traceability

In this book we recognize *verifiable digital credentials (VDCs)* as data-rich, socio-technical systems of assets and infrastructures that embody particular values, power, and assumptions for record checking, confirmability, and traceability.

We intentionally use *'verifiable digital credentials'* or 'verifiable credentials' rather than simply *'digital credentials,'* or *'blockchain credentials'* and are mindful of when we use the term *'verifiable'* to maintain clarity about both their socio-technical nature and mechanisms of verification.

▸ The term *'verifiable'* we use to mean the processes of how claims made can be checked, confirmed, and/or are traceable with and to the primary source, or via a trusted second or third-party. It highlights the critical questions of confirmability and trust, who has the authority to verify, what processes social or technical are used to check, confirm or trace claims made, and how verification standards are designed and deployed

▸ The term *'digital'* acknowledges that these are computational systems that exist as data processed by computers, algorithms, and networks, carrying all the inherent biases, accessibility barriers, infrastructural dependencies, and power structures embedded in digital technologies. The *'digital'* nature means these credentials require technological literacy, internet access, compatible devices, digitized data, and participation in digital ecosystems that may exclude or disadvantage people who lack these resources or who face risks of digital surveillance.

▸ The term *'credential'* we use to mean *'any artifact that can give credence or credibility to the claims made about an entity'* be that a human, animal, group, or organizational entity. This may include claims made about who someone is (such as an identity credential), who an organization or company is (such as company registration papers), what people or organizations are allowed to do (such as work, trade, or travel authorisation credentials), someone's character, experiences, and social-emotional skills (such as a recommendation letter), and the knowledge specific, skills and expertise that people have (such as a diploma, certificate, or transcript) as well as a many other lived experiences (such as a painting, sculpture, photography).

We acknowledge that through how we design, govern, and deploy them, VDCs fundamentally shape how claims about an person or entity may be checked, confirmed, and traced, who gets believed and trusted, whose claims count, how people can move through systems, and who governs the system of trust and confirmability. The promise of VDCs to enable a more granular, portable, and people-controlled systems, a credentialing system must be weighed against the risks of creating newer forms of digital exclusion or surveillance.

The expansive lens we take in this book is especially critical when considering how VDCs may impact people from justice-impacted, immigrant, disabled, and economically marginalized communities, as well as from creative, service-based, and care communities who have historically faced barriers in having their identity, experiences, knowledge, skills, and expertise validated and recognized for verifiability. This is especially important in systems designed which harm and exclude them.

In this book we use the terms: verifiable digital credentials, VDCs, or verifiable credentials when the digital nature is clear from context.

BUILDING A FUTURE CENTERED IN CARE

This book is not against LWRs issued as VDCs. It is a challenge to the extractive design practices and unchecked technological optimism that too often drives their development.

We advocate for care-centered design that prioritizes people, human dignity and community wellbeing through foundational principles of care (see Figure 2):

▶ **Prioritize the wellbeing** of those who interact with our learning and work systems

▶ **Ensuring genuine informed consent** from those whose data is collected, used, and shared

▶ **Establishing meaningful accountability** to impacted communities

▶ **Maintaining an active commitment** to justice in both design processes, deployment, and outcomes

Figure 2. Principles for Centering Care

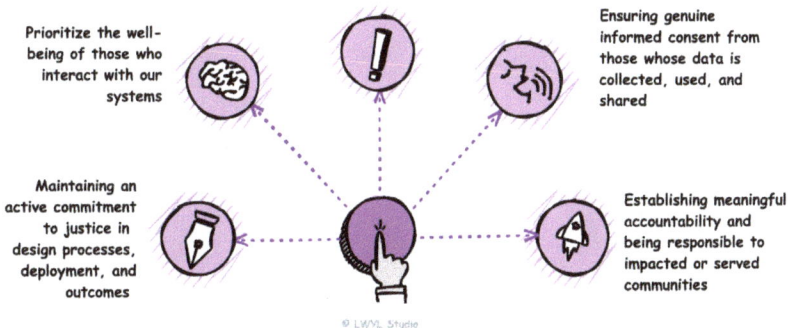

Prioritize the well-being of those who interact with our systems

Ensuring genuine informed consent from those whose data is collected, used, and shared

Maintaining an active commitment to justice in design processes, deployment, and outcomes

Establishing meaningful accountability and being responsible to impacted or served communities

© LWVL Studio

These four principles recognize that technology is never neutral and that equitable systems require intentional design choices that center the rights, experiences, and leadership of those most affected by these infrastructures. This means designing *with* rather than just for people and communities. It means treating privacy through informed consent as a fundamental right, not a luxury or add-on feature. It means measuring success not just by technical functionality, yet by whether these innovations genuinely expand opportunity without reproducing existing harms.

If we are serious about equity in learning and work systems, if we truly believe everyone deserves pathways to economic security, prosperity, and personal growth, we cannot afford to build tools that deepen systemic inequities in the name of innovation. The stakes are too high, and the communities most impacted have already waited too long for technology that truly serves rather than surveils, profiles, and harms them.

We must build with the people whose lives are most shaped and impacted by these systems and innovations.

Let us begin!

FURTHER READING

1. Benjamin, R. (2019). *Race after technology: Abolitionist tools for the new Jim code*. Polity Press.

2. Costanza-Chock, S. (2020). *Design justice: Community-led practices to build the worlds we need*. MIT Press.

3. Eubanks, V. (2018). *Automating inequality: How high-tech tools profile, police, and punish the poor*. St. Martin's Press.

4. Lum, D. (2011). *Culturally competent practice: A framework for understanding diverse groups and justice issues* (4th ed.). Brooks/Cole.

5. Noble, S. U. (2018). *Algorithms of oppression: How search engines reinforce racism*. NYU Press.

6. Page, K. L., Hansen, T., Saunders, D., Long, P., Zagidulin, D., Barfoot, R., Severs, N., & Otto, N. (2024). *LER for SBHA guide and toolkit*. U.S. Chamber of Commerce Foundation, T3 Innovation Network. https://www.t3networkhub.org/ler-toolkit

CHAPTER

01

THE HISTORY OF RECORD KEEPING, CREDENTIALING, AND STORAGE

- ▶ How record keeping and credentialing systems evolved from ancient civilizations to today

- ▶ Why institutional control over learning and work records shapes today's credential landscape

- ▶ Why record and credential design choices carry profound ethical implications for access and equity

- ▶ How storage evolution represent both opportunity and disruption in record and credential management

Learning and work record keeping and credentialing and the way we manage them is not new. The practice of documenting an individual's role, skills, learning and work activities, affiliations, and achievements, has existed since ancient times.

From engraved seals in ancient Egypt (3100 BCE - 30 BCE) to detailed employee records during the Tang Dynasty (618 - 907 CE), to handwritten university diplomas in medieval Europe (500 - 1500 CE), humans have long used records and credentials to signal completion, trust, skill, and affiliation across time and geography. As records and credentials evolved, so too has the use of systems within which we carry, store, and manage them.

Before we can really understand the risks and promises of learning and work records issued as verifiable digital credentials, we must begin with a broader story of how they have evolved, one that stretches across centuries and continents.

This chapter traces the history of the practices of learning and work record keeping credentialing, and the containers (physical, technical, as well as digital) people have used to store them. We also examine how these same records and credentials have historically been used both to recognize as well as to restrict people, and why this history must inform how we design systems in which LWRs are issued as VDCs today.

THE HISTORY OF LEARNING AND WORK RECORD KEEPING

A learning and/or work *'record'* is a *'documented account of an individual's activities, achievements, skills, or performance that serves as evidence of their capabilities, experiences, or progress.'* These records may be intentionally as well as unintentionally created and preserved to inform decisions about employment, advancement, education, or professional recognition.

To better understand the evolution of learning and work record keeping and records themselves, we conducted a historical analysis to display key events, as well as the typology, format, and purpose of record keeping across contexts.

Historical Timeline of Record Keeping

The documentation of learning and work activities, achievements, and events, dates back to ancient times (See Table 2). From early systems in Ancient Mesopotamia (3500-3200 BCE) which were used to document trade transactions and manage agricultural surplus; to Ancient Egypt (3100-3130 BCE) where elite scribes trained in hieroglyphics, serving temples and pharaonic administration to help maintain a huge centralized economic system. In Ancient China (206 BCE-220 CE) there is documented evidence of record keeping used to create a systematic government employment and promotion system, which served as a foundation for a merit-based civil service and a comprehensive personnel management system still in use today.

For centuries people, organizations, and governments have been engaged in the activity of documenting activities of learning and work, in multiple ways and for multiple purposes. As shared in Table 2 (and in Appendix B) Many of these early systems are deeply grounded in cultures of exclusivity, control, and for tracking people.

Table 2. Historical Timeline of Learning & Work Record Keeping

Period	System	Details
3500–3200 BCE	Ancient Mesopotamia (Sumerians)	Invented cuneiform script using wedge-shaped marks on clay tablets with reed stylus. Needed to document trade transactions and manage agricultural surplus. Created foundation for all future writing systems and record keeping. Complex system with over 700 symbols that changed between cities and over time.
3100–3130 BCE	Ancient Egypt	Elite scribes trained in hieroglyphics, serving temples and pharaonic administration. Maintained a centralized economic system and preserve religious and cultural values. Helped Egyptian civilization maintain a huge centralized economic system. Literacy rate was only 1% so was an extremely exclusive system.
2900–2334 BCE	Mesopotamian Scribal Schools (Eduba)	Students aged 8-22 learned mathematics, and administration in the "House of Tablets." Trained scribes for temple and palace administrative needs. Ensured literature, science, and law were transmitted across generations. Only children of upper class could afford tuition; mostly male students.
206 BCE–220 CE	Han Dynasty China	First emperor Liu Bang ordered candidates to register with character, appearance, and age recorded. Created a systematized government employment and promotion system. Established merit-based civil service. Limited to government officials only.
618–907 CE	Tang Dynasty China	Detailed employee files recording personal information, work experience, job performance, and references. Maintained efficient government bureaucracy and prevent corruption. Created comprehensive personnel management system still used today. Files could control major life decisions like marriage and travel.
11th–16th Century	Medieval European Guilds	Apprentice (7 years) → Journeyman → Master system with written agreements and masterpieces. Protected industry from competition; maintain quality standards; increase political influence. Mutual aid, quality control, professional mobility, education support. Became hereditary and exclusionary; set artificially high standards.
17th–18th Century	English Statute of Artificers and English Apprenticeship Records	First national apprenticeship system; 7-year terms, maximum 3 apprentices per master. Standardized training and prevent exploitation. Government revenue and legal protection. Created legal framework and system for professional training. Only applied to trades that existed when law was passed. Many informal apprenticeships avoided the system to avoid taxes.
19th–20th Century	Industrial Transition	Official records kept when stamp duty was payable on indentures of apprenticeship. Guild power faded due to industrialization and rise of nation-states. Mass production required standardized methods rather than guild secrets. Enabled industrial scale production and innovation. Many former handicraft workers forced into manufacturing with less job security.

© LWYL Studio

Record Typology

A typology of learning and work records emerged from the historical analysis showing the different types of record keeping and records used across the centuries (see Table 3, and Appendix C).

Interestingly, while the format may have changed (e.g., clay tablets to paper, to digital and data systems), and the social structures and technologies supporting them have evolved dramatically, the types have remained remarkably consistent.

Table 3. Typology of Learning and Work Records

Type	Historical Version	Modern Version
Student Progress Records	Mesopotamian edubba tablets showing cuneiform writing exercises and curriculum stages	Transcripts, grade reports, learning portfolios
Professional Certification	Medieval guild masterpieces and journeyman certificates proving craft competency	Professional licenses, certifications, diplomas
Employment History	Tang Dynasty personnel files recording work experience, job performance assessments, and references	Résumés, employment records, performance reviews
Apprenticeship Agreements	English apprenticeship indentures documenting terms, duration, and completion	Internship agreements, training contracts

Payroll and Wage Records	Mesopotamian clay tablets documenting worker wages, obligations, and rights between employers and workers	Payroll records, employment contracts
Educational Enrollment	Edubba school admission records for children of upper class and nobility	School enrollment records, admissions files
Skill Assessment	Sumerian curriculum records tracking progression through Tetrad and Decad compositions	Competency assessments, skill certifications
Guild Membership	Medieval guild rolls documenting member status, dues, and privileges	Professional association membership
Disciplinary Records	Sumerian "Schooldays" describing student punishments and behavioral issues	Disciplinary files, conduct records
Teaching Credentials	Mesopotamian scribal qualifications for temple and palace instruction roles	Teaching licenses, faculty credentials
Trade Secrets and/or Knowledge	Medieval guild records of protected techniques and methods	Proprietary training materials, trade documentation
Social and/ or Political Background	Chinese dang'an files including political affiliations and family background	Background checks, security clearances
© LWYL Studio		

Record Format

The format and design of learning and work records has also changed over the centuries (Table 4).

Table 4. Formats of Learning and Work Records

Era	Medium	Content	Access
Ancient	Clay tablets, stone	Basic transactions, names, quantities	Very limited to the scribal elite only
Classical	Papyrus, parchment	Personal details, performance assessments	Restricted to officials
Medieval	Paper, vellum	Formal agreements, guild membership	Controlled by organizational institutions and governments.
Modern	Paper, digital, verifiable	Comprehensive documentation from across learning and work contexts	Only accessible to people with digital literacies and devices, and multiple digital formats and limited interoperability makes it more difficult to access and use across systems and contexts.

Record Keeping Purpose

Over the centuries, learning and work records have been mostly used to serve four main purposes: *Credentialing, Administrative Control, Social Mobility, and Quality Assurance* (See Table 5).

Table 5. Purpose of Learning and Work Record Keeping

Credentialing	Administrative Control
▶ Prove qualification for roles or advancement ▶ Demonstrate completion of required training ▶ Validate specific skills or knowledge	▶ Track progress through systems ▶ Manage resource allocation ▶ Monitor compliance with standards
Social Mobility	**Quality Assurance**
▶ Enable movement between positions or locations ▶ Provide evidence for advancement opportunities ▶ Create portable proof of capabilities	▶ Maintain professional standards ▶ Document adherence to established practices ▶ Enable peer review and validation

EARLY FORMS OF CREDENTIALS

The word 'credential' comes from the Latin *credere*, meaning 'to believe' or 'to trust.' Historically, credentials were coded physical tokens, letters, or seals that conveyed trust from one party to another of a type of record.

Ancient Civilizations

In Ancient Egypt (circa 3000 BCE), skilled workers and artisans in royal service carried inscribed stones or seals to show their authorized role. These served as early status markers. In China's Tang Dynasty (7th century CE), the Imperial Examination system required students to pass standardized tests to gain bureaucratic roles. Successful candidates received certificates that acted as state-sanctioned credentials. Roman soldiers and messengers carried bronze diplomas, documents confirming their rights, often issued upon discharge from service.

Medieval Europe

Early credentials were fragile, region-specific, and prone to forgery. Yet, they enabled mobility, especially for artisans, scholars, and officials in pre-modern economies. Guilds issued membership tokens, letters of apprenticeship, or *freemen's papers* to skilled laborers. Monasteries and universities provided handwritten letters (*litterae testimonials*) to confirm completion of theological or scholarly studies. These evolved into degrees and transcripts.

Over time, credentialing became more formalized through institutions. As education and employment systems formalized, so did documentation, transcripts, licenses, apprenticeships, and references became common mechanisms to verify ability and control access to opportunity.

Universities and Diplomas

The University of Bologna (est. 1088) and University of Paris (est. 1150) issued parchment diplomas to graduates, some of the earliest standardized education credentials. By the 14th century, European universities created registries of students, courses, and exam outcomes. These paper records laid the foundation for the modern transcript.

Employment Records

In Elizabethan England, the 1563 Statute of Artificers required written contracts and service records for apprenticeships. Industrial-era employers (1800s) kept personnel files and began issuing letters of reference, precursors to modern job records. In the 20th century, the rise of HR departments led to systematic internal documentation of training, evaluations, and promotions.

These records increasingly became tools not just for tracking, yet for verifying and controlling worker access to opportunities.

The Rise of Standardized Credentials

In the 20th century, credentials became central to modern education and labor systems.

The transcript emerged as a formalized tool in U.S. higher education in the early 1900s. Diplomas, licenses, and certificates expanded in health, law, and technical fields to signal professional competency. The General Educational Development (GED) test (1942) offered high school equivalency certification, making credentials available to more diverse learners.

By the 1980s, machine-readable student records enabled computer-based credentialing in universities. Credentials became increasingly bureaucratic and required for access to formal employment, *especially* for people from marginalized communities.

Over the years credentials become a cornerstone of learning and work records (LWRs) because they serve as the *verified, portable evidence* of someone's learning and capabilities.

THE FIRST KNOWN 'WALLET' FOR CREDENTIALS

With the rise in the number and format of learning and work records, including credentials has come the need for a personal storage system to hold, and manage all your records. Their relatedness could be considered and differentiated based on what each provides (Figure 3).

Figure 3. Records, Credentials, Wallets

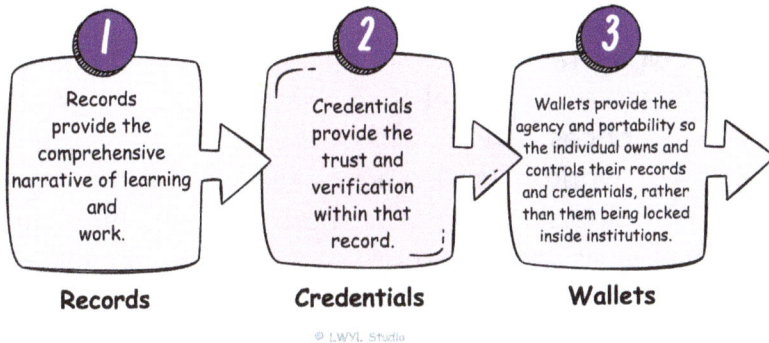

① Records provide the comprehensive narrative of learning and work.

Records

② Credentials provide the trust and verification within that record.

Credentials

③ Wallets provide the agency and portability so the individual owns and controls their records and credentials, rather than them being locked inside institutions.

Wallets

© LWYL Studio

- ▶ A **Record** provides the **comprehensive narrative** of learning and work.

- ▶ **Credentials** provide the **trust and verification** within that record.

- ▶ A **storage system like a wallet** can provide the **agency and portability** so an individual can own and control their records and credentials, rather than them being locked inside institutions. These may be called wallet, backpacks, portfolios etc.

Where did the idea of a wallet come from

The ancient Greek word *kibisis*, which is said to describe the pouch carried by the god Hermes and the sack in which the mythical hero Perseus carried the severed head of Medusa, has been typically translated as *"wallet"*.

In the late 14th century the word 'wallet' was used meaning "bag" or "knapsack," from the uncertain origin (Norman-French golette (little snout)?), or from a similar Germanic word, from the Proto-Germanic term "wall," which means "roll" (from the root "wel," meaning "to turn or revolve."

Shakespeare's early usage described something that we would recognize as more like a backpack today. The modern meaning of "flat case for carrying paper money" was first recorded in 1834 in American English.

As shown in Table 6., *The History of Wallets*, the idea of a wallet to store credentials has analog roots: Paper portfolios: Apprentices, artists, and tradespeople often carried portfolios or workbooks to show examples of their skills. Document holders: Immigrants and travelers carried paper folders with visas, identification, and employment letters.

Table 6. The History of Wallets

Period	Civilization	Wallet	Description
Ancient Times	Early Humans	Animal Skin Pouches	The concept of carrying tools, food, and valuable items in pouches dates back to ancient times. Early humans created simple pouches using animal skins or cloth tied with strings
	Ancient Greece & Rome	Leather Pouches	Greeks and Romans carried coins in small leather pouches, often tied to their belts. These pouches were known as "Kibisis" and "Bursa."
Medieval Period	Medieval Europe	Girdle Puses	During the medieval period, wallets evolved into "girdle purses," small bags attached to belts. They were used to carry coins, herbs, and personal items.
		Secure Pouches	Pickpocketing was common, leading to the creation of more secure designs, sometimes with metal clasps
Renaissance and Enlightenment	15th to 18th Century	Paper Currency	The introduction of paper currency in Europe led to the need for more specialized wallets to carry notes. Early wallets were designed to hold documents, coins, and personal identification.
		Billfolds	By the 17th century, wallets had evolved into "billfolds," which were flat and designed to carry paper currency and identification cards.

Industrial Revolution	19th Century	Mass Production	The Industrial Revolution enabled the mass production of wallets, making them more accessible to the general public.
		Design Innovations	Wallet designs are diversified, incorporating various compartments for coins, notes, and later, photographs.
20th Century to Present	20th Century	Credit Cards	The introduction of credit cards in the 1950s revolutionized wallet design, necessitating slots specifically for cards.
		Materials	Wallets began to be made from various materials, including leather, nylon, and synthetic fabrics.
	21st Century	Digital Accounts such as Wallets	The rise of digital payments and smartphones led to the development of digital accounts called wallets (e.g., Apple Pay, Google Wallet), which store payment information electronically.
		Cryptographic Accounts or Wallets	The creation of digital wallets to hold cryptographic assets, commonly known as crypto wallets, emerged from the development of blockchain technology and the launch of Bitcoin in 2009. These wallets are essential tools in the decentralized digital economy, allowing individuals to store, manage, and use cryptographic keys to access and control their digital assets securely.

© LWYL Studio

Digital Wallets: The First Known Instance

The digital wallet is a modern evolution of old practices: Physical portfolios and identity documents eventually gave way to early federated ID systems and, more recently, digital wallets based on decentralized standards like DIDs and verifiable credentials.

- The Liberty Alliance Project (2001) aimed to create federated identity systems with secure, portable credentials.

- In 2005, Microsoft's InfoCard tried to offer a personal identity selector, a digital wallet to hold verifiable credentials.

- The W3C Verifiable Credentials Data Model (2018) and Decentralized Identifiers (DIDs) introduced the technical standards that power today's digital wallets.

- Tools like uPort, Sovrin, and later Truvera, and DIF Wallets represent the first digital wallets purpose-built for verifiable credentials.

These tools enable people to hold, carry, and manage credentials issued by institutions and present them to verifiers, with cryptographic trust often built in.

RECORDS AND CREDENTIALS AS TOOLS OF HARM

While records of learning and work and credentials, have enabled opportunity, they have also caused harm.

Throughout history, credentials have been used to exclude, control, and surveil, disenfranchising marginalized people through racist passbooks, gendered gatekeeping, and digital surveillance, showing that design choices carry profound ethical stakes.

Racial and Colonial Exclusion

In colonial South Africa, passbooks functioned as internal passports used to control Black workers' movement and employment. In the United States, Jim Crow-era voter literacy tests required credential-like documentation, denying Black Americans access to democratic participation.

Gender, Immigrant, and Economic Gatekeeping

Women were excluded from apprenticeships, universities, and professions, as well as owning property and bank accounts that required credentials well into the 20th century. Licensing bodies in fields like law or medicine often used credentials to keep out immigrants and working-class entrants.

Policing and Surveillance

Germany in the lead up to, and during World War 2 used work permits and identity documents to track and persecute Jewish citizens, gypsies, immigrants, and many others. The U.S. Patriot Act enables digital credential checks that disproportionately target Muslim communities and immigrants from the Middle East.

Digital Harm

In modern settings, credential platforms have been used to:

- ▶ Deny jobs based on juvenile or criminal records.
- ▶ Revoke access to education due to administrative holds.
- ▶ Share private educational history with third-party advertisers, technology vendors, and data collection agencies.

Records of learning and work, including many types of credentials are not neutral, they are shaped by power and the social design of exclusion (inclusion).

LESSONS FROM HISTORY

Lessons from history remind us that the systems we build today will shape lives for generations, making care and responsibility essential. So what have we learned?

1. **Credentialing is always political.** It decides who is 'qualified,' who is 'deserving,' and who is 'visible'.

2. **Record and credential systems can both liberate and confine.** A GED may offer a second chance, while a professional license requirement may serve as a barrier.

3. **Portability has always mattered.** Whether in a leather folio or a digital wallet, people need control AND ownership over how they carry and share their records.

4. **Systems must be designed with care.** Technologies that ignore history risk repeating its harms.

CONCLUSION

The history of credentials is complex and contradictory. They have served as pathways to recognition, and as mechanisms of exclusion. From ancient seals to digital diplomas, the purpose and power of credentials has always depended on who controls them, who recognizes them, and who benefits from their use.

As we build digital credentialing systems today, we must remember this history. Because if we don't, we risk building a future where credentials replicate past injustices, only faster, and at scale.

FURTHER READING

1. Brown, J. S., & Duguid, P. (2000). *The social life of information*. Harvard Business School Press.

2. Burke, P. (2000). *A social history of knowledge: From Gutenberg to Diderot*. Polity Press.

3. Collins, R. (2019). *The credential society: An historical sociology of education and stratification* (Legacy ed.). Columbia University Press. (Original work published 1979)

4. Cottom, T. M. (2017). *Lower ed: The troubling rise of for-profit colleges in the new economy*. The New Press.

5. Eubanks, V. (2018). *Automating inequality: How high-tech tools profile, police, and punish the poor*. St. Martin's Press.

6. Foucault, M. (1977). *Discipline and punish: The birth of the prison* (A. Sheridan, Trans.). Vintage Books. (Original work published 1975)

7. Pew Research Center. (2016, September 20). *Digital readiness gaps*. https://www.pewresearch.org/internet/2016/09/20/digital-readiness-gaps/

8. Pew Research Center. (2021, June 22). *Digital divide persists even as Americans with lower incomes make gains in tech adoption*. https://www.pewresearch.org/short-reads/2021/06/22/digital-divide-persists-even-as-americans-with-lower-incomes-make-gains-in-tech-adoption/

9. W3C. (2022, March 3). *Verifiable credentials data model v1.1*. https://www.w3.org/TR/vc-data-model/

10. Yeo, G. (2021). *Record-making and record-keeping in early societies*. Routledge. https://doi.org/10.4324/9780429054686

CHAPTER

02

CREDENTIAL FRAMEWORKS: GLOBAL STANDARDS OR LOCAL STRICTURES?

- ▸ Frameworks prioritize standardization over inclusion
- ▸ Most frameworks are not built with or for learners and workers
- ▸ Level-based hierarchies can reinforce inequity
- ▸ Ethical redesign is possible, yet must be cocreated and/or codesigned

In the rapidly evolving world of digital credentialing, a foundational layer often goes unnoticed by most learners, workers, and even many designers: credential frameworks.

These are the taxonomies, hierarchies, and standardization systems that shape what can be recognized, how it is validated, and where it fits in an ecosystem of meaning and mobility. At their best, frameworks help align qualifications, support learner mobility, and provide transparency. At their worst, they enforce narrow definitions of value, embed Western-centric assumptions, and create inaccessible bureaucracies that fail to serve the people they claim to benefit.

In this chapter we examine international credential frameworks, and explore how these systems influence global and domestic record and credentialing efforts. We analyze their design, accessibility, and implications, and argue for a rethinking of how frameworks are constructed and for whom they serve.

WHAT ARE CREDENTIAL FRAMEWORKS?

Credential frameworks are structured systems for categorizing, comparing, and communicating learning achievements. Most frameworks are designed to:

- ▸ Establish levels of learning or competence
- ▸ Define types of qualifications or credentials

- Map how different learning outcomes relate
- Facilitate mobility across borders or institutions

Frameworks may include:

- European Qualifications Framework (EQF) across European Countries
- National Qualification Frameworks (NQFs) in countries like South Africa, Australia, the UK, and Malaysia
- Credential Transparency Description Language (CTDL) in the U.S., managed by Credential Engine
- ASEAN Qualifications Reference Framework
- UNESCO conventions on recognition of qualifications

WHO DESIGNS THESE FRAMEWORKS?

Frameworks are typically created by policymakers, academic bodies, international agencies, and consultants, often removed from the lived realities of learners and workers. They reflect:

- Institutional logics, formal, academic, degree-based hierarchies
- Western educational models, linear progression, individual achievement
- Economic priorities, labor market alignment over personal development

Voices absent in design often include:

- Indigenous leaders and knowledge holders
- Informal and social sector workers
- Adult learners, reentrants, migrants, refugees
- Learners and workers from marginalized and historically excluded communities

▸ Learners in the global south or emerging economies

In many cases, credential frameworks are imposed without meaningful participation from those most impacted by their classifications and design.

WHO CAN USE THESE FRAMEWORKS?

Qualification frameworks are typically designed for institutional actors, governments coordinating national systems, educational institutions mapping programs, technology vendors building platforms, rather than for learners and workers navigating their own pathways. This design orientation creates significant accessibility challenges:

▸ **Language barriers**: Technical terminology and institutional jargon limit comprehension

▸ **Structural complexity**: Schemas and ontologies prioritize system interoperability over user understanding

▸ **Cultural distance**: Limited translation and minimal adaptation to diverse contexts

▸ **Narrative absence**: No provision for learners to contextualize their experiences or challenge categories

The Credential Transparency Description Language (CTDL) illustrates these tensions. Developed to standardize credential metadata across systems, CTDL enables the data infrastructure necessary for transparency and portability. However, its architecture was built for systems engineers and technical implementers, not for learners seeking to understand their options or for community organizations supporting non-traditional populations. The schema requires familiarity with linked data concepts, JSON-LD serialization, and controlled vocabularies, technical knowledge rarely possessed by those most affected by credentialing decisions. Moreover, CTDL's emphasis on machine-readable structure leaves little room for the qualitative dimensions of learning: the struggle to master a skill, the community context that gave a credential meaning, or the barriers overcome in its pursuit.

The European Qualifications Framework (EQF) raises parallel concerns. Its eight-level structure uses descriptors, knowledge, skills, autonomy, and responsibility, to function as a 'translation grid' enabling qualification comparisons across European countries. While this supports mobility and recognition, the descriptors themselves embed particular cultural assumptions. Terms like 'autonomy' and 'responsibility' are not culturally neutral; they reflect specific traditions of individualism and professional identity that may not align with collectivist cultures, Indigenous knowledge systems, or different models of professional practice. Without mechanisms for culturally responsive interpretation, such frameworks risk privileging certain forms of achievement while marginalizing others.

THE INFLEXIBILITY OF LEVELS

Credential frameworks often impose rigid *levels* or hierarchies such as 'Level 1 through Level 8', or 'low-skill to high-skill' or 'beginner to advanced' Yet, in this reductionist approach of categorization and approach we are inflexible to:

▶ **Skills and abilities** as not always hierarchical, yet always contextual and intersectional

▶ **Knowledge and wisdom** can be relational, experiential, or collective, not just individual; as well as declarative, procedural, and episodic

▶ **Communities** may not agree on what counts as more or less specialised, advanced, and/or masterly

▶ **Many people** from marginalized communities can demonstrate competence in domains that often lack formal validation or recognition (e.g., caregiving, organizing, community and social leadership, cultural and indigenous ways of knowing).

By forcing all recognition into predefined levels, frameworks and taxonomies we more often devalue pluralistic and more social forms of learning, knowing, and being.

CULTURAL NON-RESPONSIVENESS

Beneath their technical neutrality, qualification frameworks encode specific cultural assumptions about the nature and validation of knowledge:

▶ **Privileging formal education**: Learning is legitimate when it occurs in recognized institutions following documented curricula, while apprenticeship, elder transmission, and community-based learning are categorized as 'informal' or 'non-formal, or alternative, terms that already signal lesser status

▶ **Universalizing Western formats**: Diplomas, degrees, and certificates become the template for all credentials, making other forms of recognition such as ceremonial authorization, lineage-based transmission, community endorsement, illegible to the system

▶ **Reducing transferability to market value**: Credentials are assessed primarily by their portability across employers and borders, not by their significance within cultural, spiritual, or community contexts

These design choices systematically exclude entire knowledge traditions. Indigenous learning systems often embed knowledge in ceremony, relationship, and land-based practice, dimensions that cannot be captured in learning outcome statements or competency frameworks. Community validation mechanisms, such as the recognition a traditional healer earns through decades of practice and community trust, resist translation into the language of credit hours and qualification levels. Forms of learning that are inherently collective, relational, or spiritual, where knowledge belongs to a community rather than an individual, or where understanding is inseparable from specific relationships and contexts, find no accommodation in frameworks designed around individual achievement and abstract, decontextualized skills.

"This pattern exemplifies what Miranda Fricker (2007) calls epistemic injustice: the structural marginalization of certain ways of knowing such that entire worldviews are excluded from legitimacy."

When frameworks claim comprehensiveness while operationalizing only Western educational paradigms, they don't simply overlook diverse knowledge systems, they actively construct them as lesser, supplementary, or illegitimate, reinforcing colonial hierarchies of knowledge.

CASE EXAMPLE: AFRICAN CONTINENTAL QUALIFICATIONS FRAMEWORK

The African Continental Qualifications Framework (ACQF), validated in 2023 after four years of development, demonstrates both the promise and limitations of regional qualification coordination. As Commissioner Mohamed Belhocine of the African Union stated at its launch, the ACQF aims to ensure *'knowledge, skills, and competencies attained by our learners are valued and recognized not only within individual countries but across borders,'* supporting mobility in a continent where 10-12 million youth enter labor markets annually but only three million formal jobs exist.

The development process involved extensive consultation across over 40 African countries, producing a policy document, ten technical guidelines, and comprehensive capacity-building materials. Yet several tensions have emerged that illuminate broader challenges in qualification standardization:

▶ **Structural alignment versus contextual fit**: The ACQF adopts an eight-level structure resembling the European Qualifications Framework, enabling comparison with European credentials, a clear benefit for mobility and recognition. However, this design choice imports assumptions about linear knowledge progression that may not reflect African educational realities. As researchers note, when frameworks align with 'global templates,' they risk making 'the specific national context largely irrelevant.'

▶ **Formal systems versus lived reality**: The vast majority of African workers develop skills outside formal institutions, through apprenticeships, agricultural knowledge passed through families, trading skills learned in markets, craft knowledge transmitted in workshops. While the ACQF acknowledges Recognition of Prior Learning, its competency-based structure assumes documented, standardized assessment. A master weaver in Burkina Faso or an expert mechanic in Nairobi's Gikomba market may possess sophisticated knowledge that resists translation into level descriptors and learning outcomes.

▶ **Diverse knowledge systems**: From centuries-old Islamic educational traditions in the Sahel to Indigenous knowledge systems embedded in specific languages and territories, from oral histories deliberately kept outside written form to healing practices validated through community recognition rather than certification, African learning encompasses epistemologies that qualification levels cannot easily accommodate. When asked how such knowledge might be included, framework developers often default to 'documenting' and 'assessing' it, processes that may fundamentally transform or diminish what they claim to recognize.

▶ **Implementation constraints**: National officials implementing the ACQF report navigating competing pressures: donor requirements for standardized reporting, software platforms designed for European systems that resist adaptation, and insistence on 'global comparability' that effectively means conformity with Northern frameworks. French-speaking countries have been particularly vocal in requesting support for developing nationally appropriate approaches rather than adopting imported structures wholesale.

The ACQF's trajectory raises essential questions: Can a continental framework genuinely serve 55 diverse nations and countless knowledge traditions? Or do the structural requirements of cross-border recognition such as standardization, documentation, comparability, inevitably privilege certain forms of learning while marginalizing others? Implementation itself may reveal whether the framework serves African learners and workers, or primarily serves the administrative needs of institutions and the mobility requirements of a narrow segment of formally educated professionals with the means to travel across international borders. .

WHO BENEFITS FROM THE FRAMEWORKS?

The architecture and adoption of qualification frameworks reflect the interests of their primary interested parties or partners in the development, which are rarely the learners and workers frameworks claim to serve. The frameworks largely benefit the following groups and organizations.

1. **Governmental priorities**: National governments promote frameworks to advance workforce planning, demonstrate education system quality in international comparisons, and manage immigration through credential-based sorting. *These applications serve state administrative needs, though their relationship to actual labor market outcomes or learner success remains contested in research.*

2. **Donor and development agendas**: International organizations and philanthropic funders favor qualification frameworks because they generate measurable outcomes and comparable data. This creates self-reinforcing demand: *donors fund framework development because frameworks produce the standardized metrics donors require for reporting, regardless of whether those metrics capture meaningful dimensions of learning.*

3. **Technology vendor interests**: Education technology companies, credentialing platforms, and HR software providers require structured data to function, making them natural advocates for framework adoption. Once vendors build products around particular standards, they create lock-in effects: *institutions must maintain compatibility with dominant platforms, giving vendors influence over framework evolution to align with their technical architectures.*

While frameworks are presented as serving learner mobility and credential transparency, their primary utility flows to employers conducting screening, platform developers selling services, institutions seeking accreditation, and governments generating statistics. The frameworks' complexity and institutional focus frequently make them more barrier than bridge for the workers and learners they ostensibly serve.

The Burden of Legibility

Individuals whose learning occurred outside formal systems face what might be termed a translation mandate. To be recognized, they must renarrate their knowledge in the framework's vocabulary: identify possessed 'competencies,' map experience to qualification levels, provide documentation the system accepts as evidence. This requirement places the entire burden of proof on individuals while privileging forms of learning that generate institutional documentation.

For example, a garment worker with sophisticated understanding of textile properties, production efficiency, and quality control may struggle to translate that expertise into 'Level 3 competencies in manufacturing,' not because the knowledge is insufficient but because it was developed through practice and exists in tacit, embodied forms resistant to articulation in competency statements.

The framework cannot recognize what it wasn't designed to see.

Those unable to perform this translation aren't simply disadvantaged; they're rendered invisible while the framework claims to be creating transparency.

THE MYTH OF NEUTRALITY

In summary, frameworks are not neutral. They are built on decisions about:

- ▶ What counts as knowledge or a skill?
- ▶ Who has the authority to validate it?
- ▶ What is considered 'progress' or 'advancement'?

For example, the difference between 'communication' and 'oral storytelling' is not just linguistic, it reflects which cultural practices are seen as valid skills. In this way, credential frameworks can replicate colonial logics, validating Global North forms of knowledge while erasing local epistemologies.

TOWARD ETHICAL FRAMEWORKS

Redesigning credential frameworks to serve justice rather than merely efficiency requires specific shifts in both technical architecture and power relations. Here are some was to make this shift in practice:

Participatory design as requirement, not gesture: Framework development must center those whose learning has been historically excluded, not through consultation but through shared authority. This means compensating community members as co-designers, ensuring decision-making power over how their knowledge is represented, and building accountability to affected communities.

> *The concept of "nothing about us without us" offers a model: no framework element describing Indigenous or a communities knowledge should proceed without the explicit consent and design participation of relevant Indigenous and lived communities.*

Accommodate narrative alongside taxonomy: Ethical frameworks must allow learners to present portfolios, narratives, and multimedia documentation that reveal the situated nature of their expertise. A community organizer's credential might include campaign narratives and community testimonials; a traditional healer's recognition might incorporate oral histories and apprenticeship relationships. These aren't supplementary, they're alternative forms of documentation with equal validity.

Recognize reproductive, relational, and community knowledge: Explicitly validate caregiving, community organizing, land-based learning, and other forms of labor sustaining social life. This requires moving beyond categorizing such knowledge as 'informal' or 'soft skills' and developing frameworks that can represent collective expertise and knowledge embedded in specific communities.

Support multiple recognition systems: Enable different communities to define their own pathways and markers of expertise. This may limit comparability across systems, yet preserving cultural integrity in how knowledge is recognized may be more valuable than universal commensurability.

Prioritize human legibility: Design interfaces, language, and interaction models for the people frameworks describe, not for institutions processing their data. This means multiple language options, varied literacy support, multimedia alternatives, and transparency about data use.

Create democratic governance structures: Establish ongoing participatory mechanisms for framework oversight, sustained involvement of learners and workers from diverse communities in decisions about inclusion, representation, and revision. Build clear processes for challenging framework elements that misrepresent or exclude particular forms of knowledge.

These principles shift the fundamental question from 'how do we efficiently categorize learning?' to 'how do we recognize the full range of human knowledge and capability?'

The Blackfeet Nation's partnership with Accelerate Montana demonstrates these principles in practice. Rather than forcing Blackfeet knowledge into existing credential structures, the collaboration, led by a Community Working Group from the Blackfeet Nation working with 4 Poles

Educational Consulting Group (Wendy Bremner and Lona Running Wolf), is defining nine 'living principles' as valued competencies for employment systems. These principles, rooted in Blackfeet language, stories, and philosophy, are being published to a credential registry to connect cultural identity with workplace recognition. As Wendy Bremner explains, *"When we say something like ikakimat 'trying hard,' we're not just giving a definition. We're invoking a full story, a whole philosophy."*

Critically, the Blackfeet Nation maintains authority over how these principles are defined, documented, and recognized, ensuring their epistemology isn't merely accommodated but governs how cultural knowledge is experienced, structured, and valued in employment contexts. The project also establishes cultural alignment requirements for employers operating within Tribal boundaries, embedding Blackfeet values into hiring and workplace practices rather than simply translating Indigenous knowledge into Western workforce frameworks.

CONCLUSION

Credential frameworks shape what gets seen, valued, and rewarded in digital learning systems. While they can offer clarity and alignment, they often entrench narrow, inaccessible, and culturally unresponsive models of what learning is, and who gets to be recognized.

If we want credentialing to be liberatory, not extractive, we must reimagine the foundations. That means asking not just how frameworks can be built, yet *who gets to build them, what knowledge they privilege, and whether they truly serve the people at the center of their claims.*

FURTHER READING

1. Accelerate Montana. (n.d.). *Blackfeet Living Principles as open data*. Retrieved December 19, 2024, from https://www.acceleratemt.com/blackfeetlivingprinciples

2. Allais, S. (2014). *Selling out education: National qualifications frameworks and the neglect of knowledge*. Sense Publishers.

3. Allais, S. (2017). Labour market outcomes of national qualifications frameworks in six countries. *Journal of Education and Work*, *30*(5), 457–470. https://doi.org/10.1080/13639080.2016.1243232

4. African Union. (2023). *African Continental Qualifications Framework* [Policy document]. https://acqf.africa

5. Bohlinger, S. (2012). Qualifications frameworks and learning outcomes: Challenges for Europe's lifelong learning area. *Journal of Education and Work*, *25*(3), 279–297. https://doi.org/10.1080/13639080.2012.687571

6. Bohlinger, S. (2019). Ten years after: The 'success story' of the European qualifications framework. *Journal of Education and Work*, *32*(4), 393–406. https://doi.org/10.1080/13639080.2019.1646413

7. Castel-Branco, E., & Okonkwo, C. E. (2020). *African Continental Qualifications Framework: Overview* [Infographic]. European Training Foundation. https://www.etf.europa.eu/sites/default/files/2020-07/acqf_infographic_en_2020_final.pdf

8. Credential Engine. (2023). *CTDL handbook*. https://credreg.net/ctdl/handbook

9. European Training Foundation. (2024). Rolling out the African Continental Qualifications Framework. https://www.etf.europa.eu/en/news-and-events/news/rolling-out-african-continental-qualifications-framework

10. European External Action Service. (2023, July 13). *African Continental Qualifications Framework (ACQF) is validated and phase II is launched in Addis Ababa*. https://www.eeas.europa.eu/delegations/african-union-au/african-continental-qualifications-framework-acqf-validated-and-phase-ii-launched-addis-ababa_en

11. Fricker, M. (2007). *Epistemic injustice: Power and the ethics of knowing.* Oxford University Press.

12. Raffe, D. (2013). What is the evidence for the impact of national qualifications frameworks? *Comparative Education, 49*(2), 143–162. https://doi.org/10.1080/03050068.2012.686260

13. UNESCO. (2019). *Global Convention on the Recognition of Qualifications concerning Higher Education.* https://www.unesco.org/en/legal-affairs/global-convention-recognition-qualifications-concerning-higher-education

14. Wheelahan, L., Moodie, G., & Doughney, J. (2022). Challenging the skills fetish. *British Journal of Sociology of Education, 43*(3), 475–494. https://doi.org/10.1080/01425692.2022.2045186

15. Young, M. (2005). *National qualifications frameworks: Their feasibility for effective implementation in developing countries* (Employment Working Paper No. 45). International Labour Organization.

CHAPTER

03

ISSUING LEARNING AND WORK RECORDS AS VERIFIABLE DIGITAL CREDENTIALS

- ▸ What are LWRs and VDCs?
- ▸ Socio-technical standards (W3C, OB3, SD-JWTs)
- ▸ Core components: issuers, holders, verifiers, wallets
- ▸ LER Ecosystems: data-technical, social-organizational, financial-funder ecosystems
- ▸ Current use cases: workforce, education, social benefits.

To understand how Learning and Work Records (LWRs) issued as verifiable digital credentials (VDCs) (hereafter LWRs as VCs) can cause harm, we must begin by understanding what they are, how they work, and why they have captured the attention of policymakers, technologists, and education institutions alike.

WHAT ARE LEARNING AND WORK RECORDS?

We recognize *Learning and Work Records (LWRs)* as data rich records and record infrastructures that are complex living systems.

- ▸ **The term *'learning,'*** we use to acknowledge that people develop skills, knowledge, and expertise through community knowledge-sharing, peer mentorship, lived experience, apprenticeships, self-directed study, cultural transmission, family and friend interactions, and countless other experiences that extend far beyond traditional educational structures.

- ▸ **The term *'work,'*** we use to acknowledge that people contribute value and build livelihoods through diverse experiences that extend far beyond traditional employment structures.

- ▸ **The term *'record'*** we use to mean any documentation, evidence, or trace of learning and work activities, not just formal institutional transcripts or employment histories. This may encompass official certificates and diplomas alongside community testimonials, project portfolios, peer endorsements, skill demonstrations, volunteer logs, documentation of care work, creative outputs like art, and any other artifact or asset that is a record of someone's knowledge, capabilities, or contributions. These records can be digital or physical, formal or informal, institutional or community-generated, standardized or uniquely personal. They are artifacts of one's life's activities, not merely paper, product, code or protocols to ship.

LWRs are being included as part of a growing international movement to enable skills-rich hiring and advancement activities, shifting away from proxy signals of ability and experience like degrees. The goal being to create a more accurate and inclusive representation of an individual's experiences, abilities and potential. LWRs may include many different types of information, much of which is contextual to the individual and their learning and work experience (Table 7).

Table 7. Types of Information in Learning and Work Records

Type	Description
Achievements	▸ Learning, work, or creative achievements ▸ Evaluations and/or assessments
Behavior	▸ Activity or program completions ▸ Event or experience participation
Evaluations	▸ Performance reviews ▸ References and/or recommendations
Creative	▸ Creative artifacts like sketches, plans, doodles, photographs, narration, stories, or field notes

Signals	▶ Badges, microcredentials, certificates
	▶ Licenses and industry-based certifications
	▶ Diplomas, degrees and academic-based certifications
Personally Identifiable Information	▶ Name, address, location, email address, sex and/or gender, date, place, or year of birth
	▶ Life events, stories, case studies, and/or roles

LWRs are created and self-asserted by learners and workers every moment of every day through their interactions with learning, work, and life systems. They can also be awarded, issued, and/or endorsed by schools, colleges, universities, employers, job centers, workforce boards, apprenticeship programs, bootcamps, peers, mentors, coaches, and/or peer-led learning and creative communities amongst others entities..

As creatives and collectors, people are creating, sharing, and collecting millions of learning and work records daily. In their creation and use, LWRs are ecosystem artifacts which live across multiple contexts, and systems, anchored to one or a few learners and workers.

WHAT ARE VERIFIABLE DIGITAL CREDENTIALS?

We recognize *verifiable digital credentials (VDCs)* as data-rich, socio-technical systems of assets and infrastructures that embody particular values, power, and assumptions for record checking, confirmability, and traceability.

▶ **The term 'verifiable'** we use to mean the processes of how claims made can be checked, confirmed, and/ or are traceable with and to the primary source, or via a trusted second or third-party. It highlights the critical questions of confirmability and trust, who has the authority to verify, what processes social or technical are used to check, confirm or trace claims made, and how verification standards are designed and deployed.

▶ **The term *'digital'*** acknowledges that these are computational systems that exist as data processed by computers, algorithms, and networks, carrying all the inherent biases, accessibility barriers, infrastructural dependencies, and power structures embedded in digital technologies. The *'digital'* nature means these credentials require technological literacy, internet access, compatible devices, digitized data, and participation in digital ecosystems that may exclude or disadvantage people who lack these resources or who face risks of digital surveillance.

▶ **The term *'credential'*** we use to mean '*any artifact that can give credence or credibility to the claims made about an entity*' be that a human, animal, group, or organizational entity. This may include claims made about who someone is (such as an identity credential), who an organization or company is (such as company registration papers), what people or organizations are allowed to do (such as work, trade, or travel authorisation credentials), someone's character, experiences, and social-emotional skills (such as a recommendation letter), and the knowledge specific, skills and expertise that people have (such as a diploma, certificate, or transcript) as well as a many other lived experiences (such as a painting, sculpture, photography).

Verifiable Digital Credentials (VDCs) are a digital data format defined by the World Wide Web Consortium (W3C) that allows for secure, privacy-aware, and decentralized issuance and verification of learning and work records. A VDCs can be issued with any type of LWR as well as identity, travel, creative, or any documentation. Each VDC includes a cryptographic proof of authenticity and can be verified without calling a central database (unless designed with server retrieval).

The goal of verifiable credentials is the near instant verification to make verification instant, trusted, and portable, removing the friction, delays, and misuse/abuse of paper-based systems.

THE VERIFIABLE DIGITAL LWR ECOSYSTEM

The verifiable digital LWR ecosystem (or LWR/VDC) consists of three intersecting systems: i) The data-technical system, ii) the social-organizational system, and the iii) financial-funding system.

The Data-Technical Ecosystem

The verifiable credential data-technical ecosystem includes a range of open standards and technology components (Table 8) for the creation, issuing, claiming, holding, and verifying a verifiable digital credential. This stack can be used to issue LWRs as VDCs in a secure, portable, and verifiable way; and build full digital record and credentialing systems and assets for education, employment, licensing, compliance, and more.

Table 8. Standard and Technology Components

Component	Description	Example
Unique Identifiers	Unique identifiers for issuers, holders, and verifiers that are self-sovereign.	Decentralized Identifiers (DIDs)
Data Model Specification	Defines the content and structure of a verifiable credential.	W3C VC Data Model 1.1, 2.0

Achievement Data Specification	Defines the content and structure for achievement-based credentials	Open Badges 3.0 (OB3)
Consent Specification	Consent-based specifications that enable individuals to share only a part or specific data fields of a credential.	Selective Disclosure JWTs (SD-JWTs) and Zero-Knowledge Proofs (ZKPs)
Personal Data Storage	Applications or browser-based tools that allow individuals to claim, store, and manage their credentials.	Wallets, Backpacks, Portfolio, and/or Passports
Presentation protocols	Define how VDCs are shared between parties.	CHAPI, OIDC4VC, DIDComm

THE SOCIAL-ORGANIZATIONAL ECOSYSTEM

Beyond protocols and data standards of the verifiable credential itself, issuing LWRs as VDCs are brought to life by a complex web of individuals, institutions, initiatives, and agencies who all serve as actors in the social-organizational ecosystem. Understanding this human and organizational landscape is crucial to grasping how these technologies are implemented, and where influence and power is concentrated.

The verifiable digital LWR ecosystem includes multiple actors.

▸ **The primary actors** include five groups: Learners and workers (individuals and/or groups), issuers, holders, verifiers, and system providers (Figure 4 and Table 9, 1-5).

▸ **The secondary or support actors** include four groups: service intermediaries, infrastructure stewards, policy actors, and community impact actors (Figure 4 and Table 10, A-D).

Figure 4. LWR/VDC Ecosystem - Primary and Secondary Actors

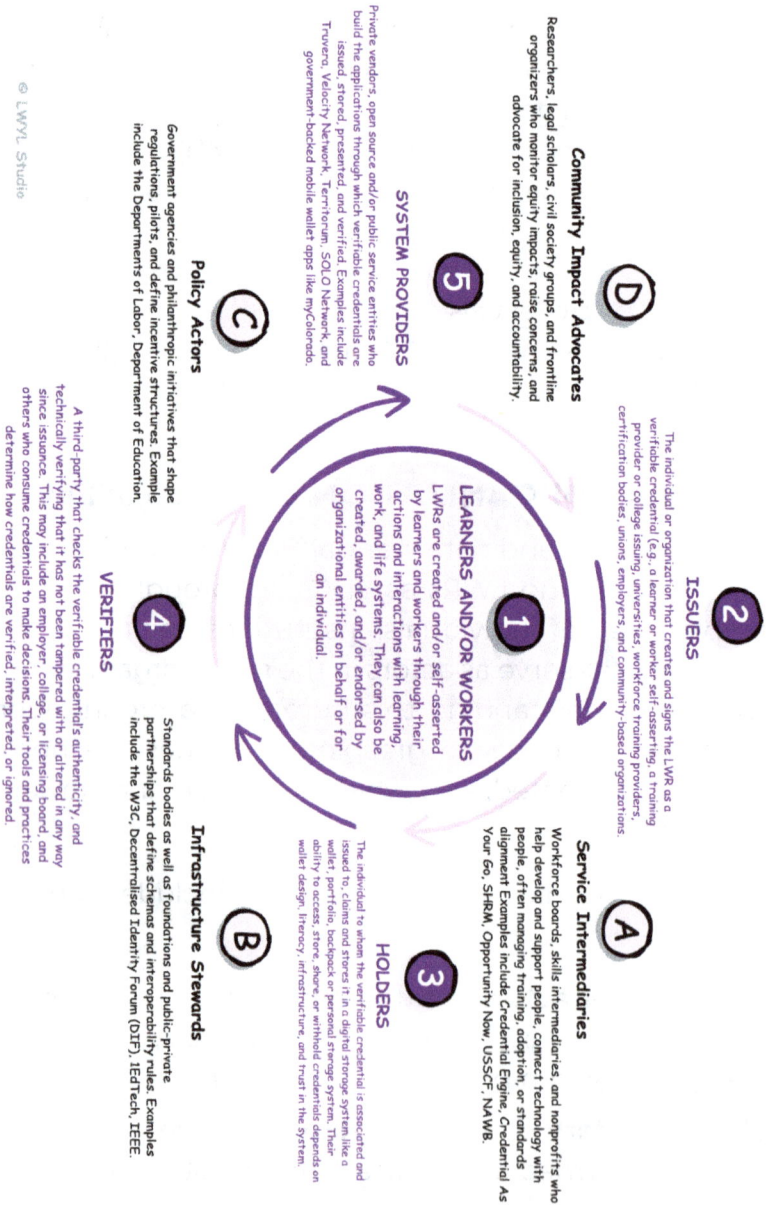

Community Impact Advocates

Researchers, legal scholars, civil society groups, and frontline organizers who monitor equity impacts, raise concerns, and advocate for inclusion, equity, and accountability.

D

SYSTEM PROVIDERS

5

Private vendors, open source and/or public service entities who build the applications through which verifiable credentials are issued, stored, presented, and verified. Examples include Truvera, Velocity Network, Territorium, SOLO Network, and government-backed mobile wallet apps like myColorado.

Policy Actors

C

Government agencies and philanthropic initiatives that shape regulations, pilots, and define incentive structures. Example include the Departments of Labor, Department of Education.

ISSUERS

2

The individual or organization that creates and signs the LWR as a verifiable credential (e.g., a learner or worker self-asserting, a training provider or college issuing, universities, workforce training providers, certification bodies, unions, employers, and community-based organizations.

1 LEARNERS AND/OR WORKERS

LWRs are created and/or self-asserted by learners and workers through their actions and interactions with learning, work, and life systems. They can also be created, awarded, and/or endorsed by organizational entities on behalf or for an individual.

VERIFIERS

4

A third-party that checks the verifiable credential's authenticity, and technically verifying that it has not been tampered with or altered in any way since issuance. This may include an employer, college, or licensing board, and others who consume credentials to make decisions. Their tools and practices determine how credentials are verified, interpreted, or ignored.

Service Intermediaries

A

Workforce boards, skills intermediaries, and nonprofits who help develop and support people, connect technology with people, often managing training, adoption, or standards alignment Examples include Credential Engine, Credential As Your Go, SHRM, Opportunity Now, USSCF, NAWB.

HOLDERS

3

The individual to whom the verifiable credential is associated and issued to, claims and stores it in a digital storage system like a wallet, portfolio, backpack or personal storage system. Their ability to access, store, share, or withhold credentials depends on wallet design, literacy, infrastructure, and trust in the system.

Infrastructure Stewards

B

Standards bodies as well as foundations and public-private partnerships that define schemas and interoperability rules. Examples include the W3C, Decentralised Identity Forum (DIF), 1EdTech, IEEE.

© LWYL Studio

WHEN CREDENTIALS CAUSE HARM

Table 9. LWR/VDC Ecosystem - Primary Actors

Actor	Description	Type
Learners and Workers	LWRs are created and/or self-asserted by learners and workers through their actions and interactions with learning, work, and life systems. They can also be created, awarded, and/or endorsed by organizational entities on behalf or for an individual.	Primary Actor - 1
Issuers	The individual or organization that creates and signs the LWR as a verifiable credential (e.g., a learner, a training provider or college issuing, universities, workforce training providers, certification bodies, unions, employers, and community-based organizations.	Primary Actor - 2
Holders	The individual to whom the verifiable credential is associated and issued to, claims and stores it in a digital storage system like a wallet, portfolio, backpack or personal storage system. Their ability to access, store, share, or withhold credentials depends on wallet design, literacy, infrastructure, and trust in the system.	Primary Actor - 3
Verifer	A third-party that checks the verifiable credential's authenticity, and technically verifies that it has not been tampered with or altered in any way since issuance. This may include an employer, college, or licensing board, and others who consume credentials to make decisions. Their tools and practices determine how credentials are verified, interpreted, or ignored.	Primary Actor - 4

System Providers	Private vendors, open source and/or public service entities who build the applications through which verifiable credentials are issued, stored, presented, and verified. Examples include *Truvera, Velocity Network, ASU Pocket, Territorum, SOLO Network*, and government-backed mobile wallet apps like *myColorado, WGU Achievement Wallet*.	Primary Actor - 5

Table 10. LWR/VDC Ecosystem - Secondary Actors

Actor	Description	Type
Service Intermediaries	Workforce boards, skills intermediaries, and nonprofits who help develop and support people, connect technology with people, often managing training, adoption, or standards alignment Examples include Credential Engine, Credential As Your Go, SHRM, Opportunity Now, USSCF, NAWB.	Secondary Actor - A
Infrastructure Steward	Standards bodies as well as foundations and public-private partnerships that define schemas and interoperability rules. Examples include the W3C, Decentralised Identity Forum (DIF), 1EdTech, IEEE.	Secondary Actor - B

Policy Actors	Government agencies and philanthropic initiatives that shape regulations, pilots, and define incentive structures. Examples include the Departments of Labor, Department of Education.	Secondary Actor - C
Community Impact Advocates	Researchers, legal scholars, civil society groups, and frontline organizers who monitor equity impacts, raise concerns, and advocate for inclusion, equity, and accountability.	Secondary Actor - D

Each of these actors brings different incentives, power, and accountability mechanisms. Technology providers and credential issuers have outsized influence in many current implementations, while holders and grassroots organizations are underrepresented.

Who gets to shape the ecosystem matters. If LWRs issued as VDCs are to be genuinely empowering, communities must not just use them; they must help define their architecture, governance, and purpose.

THE FINANCIAL-FUNDER ECOSYSTEM

Behind the scenes of every LWR and/or VDC project is a financial ecosystem that fuels its development, shapes its trajectory, and often influences whose interests are prioritized. Understanding who funds a LWR/VDC implementations and how is key to understanding the power dynamics embedded in the technology and surrounding support systems. Let's now look at each briefly.

Private Philanthropic Funders

Several large philanthropic organizations have invested heavily in LWR, VDC, and skills-based demonstrations and ecosystems. Among the most influential are:

- **Walmart.org**: Through initiatives like SkillsFWD, Reimagine Retail and the Walmart.org Center for Racial Equity, the foundation has backed workforce intermediaries, pilots, and skills-based systems focusing on workers and employers, and investing millions of dollars in the system providers who are building the solutions for testing and demonstration.

- **Bill & Melinda Gates Foundation**: A major funder of digital learning, postsecondary success, pathways, and workforce readiness programs. Gates' funding has supported research, infrastructure pilots, and policy advocacy that advance the use of LWRs in education and employment systems.

- **Lumina Foundation**: A key supporter of postsecondary credentialing innovation, Lumina funds intermediaries, research institutes, and tech pilots focusing on equitable outcomes and credential transparency.

- **Siegel Family Endowment, Carnegie Corporation, and Ascendium Education Group** have also contributed to various digital credentialing initiatives, often through coalitions like SkillsFWD or the Open Skills Network.

Investor-Backed and For-Profit Tech Vendors

Many system providers building the infrastructure for LWRs and VDCs are privately held, for-profit startups or scaleups backed by private equity. These include, yet is not limited to: *Truvera (formerly Dock.io), Velocity Network Foundation, Credly (acquired by Pearson), Badgr (acquired by Canvas*

Instructure), Parchment, iQ4, SpruceID, Territorium, and WeLibrary LLC (owner LearnCard).

While many of these companies contribute to open standards, their revenue models rely on government and enterprise contracts, platform integrations, value-added analytics and/or artificial intelligence. Their financial survival depends on adoption at scale, which can create pressure to optimize for implementation speed and financial equity demands, rather than accessibility, usability, and learners and worker needs.

Public Sector Investment

In the USA, federal and state governments also play a significant role in shaping the LWR and VDC ecosystem. Key investments include:

- **Federal Department of Labor:** Through grants to workforce boards, apprenticeships, and job quality pilots that rely on LWRs and digital skills records.

- **Federal Department of Education:** Particularly in post-secondary reform and interoperable learning records.

- **State-led initiatives** like Colorado's myColorado Digital ID Wallet, California's Cradle-to-Career Data System, Indiana's Achievement Wallet, Alabama's Talent Triad, and Credential Engine State partnerships.

Internationally, national governments and regional bodies are playing a substantial role in the development and deployment of digital learning and work records (LWRs) and digital credentials. These include (yet are not limited to):

- **Government-backed platforms**: Many countries are launching or piloting digital credential systems under public administration or in collaboration with public universities and ministries.

- ▸ **Alignment with national qualifications frameworks**: Many implementations reference established frameworks (e.g., EQF, AQF) to promote cross-institutional and international recognition.

- ▸ **Focus on verification and portability**: Common goals include preventing fraud, improving credential sharing, and supporting lifelong learning pathways.

- ▸ **Pilots and partnerships with vendors**: Several national platforms use third-party providers (e.g., Digitary, Evernym, GovTech) under government contracts or partnerships.

Public funding often flows through third-party intermediaries or is matched by philanthropic dollars in 'braided funding' models. These funds are sometimes tied to performance metrics or outcomes that further shape what systems are built and which populations they serve, yet not always creating an alignment between funding, use, and impact.

Influence on Ecosystem Design

Where funding comes from often dictates the priorities of the ecosystem as well as a the design of the solutions:

- ▸ **Private foundations** may emphasize innovation, scale, or economic mobility.

- ▸ **Venture capital** demands rapid adoption and monetization pathways.

- ▸ **Public funders** may prioritize compliance, credential transparency, or labor market alignment.

Too often, the people most affected by these systems, learners and workers are the least represented in the rooms where funding decisions are made. Without shared governance and public accountability, the financial ecosystem can drive record and credentialing infrastructure that serves more institutional and organizational business goals over community, learner, or worker needs creating a huge gap between where learners and workers are at, and what is being built.

WHY IS THIS ALL SO EXCITING?

The promise of digital records of learning and work issued as verifiable digital credentials lies in their possibility to:

▶ Give individuals more control over their learning and work records, credentials, and the data and information they include

▶ Enable rapid verification of achievements

▶ Support lifelong learning across multiple institutions

▶ Reduce hiring friction and improve equity in the job market

▶ Replace outdated resume, transcript, and diploma systems

If done well, these credentials are proposed to help level the playing field for people who've been excluded from many education and career pathways. They could also move move and raise a lot of financial capital for companies along the way.

YET ... WHAT COULD GO WRONG?

This book is about the flip side. LWRs issued as VDCs are only as ethical and impactful as the people and systems that govern them and the people and vendors who design, build, and deploy them. Within these contexts, limited understanding of the harms and/or poor implementation can lead to:

- ▶ Surveillance and over-disclosure of sensitive information
- ▶ Discrimination based on how credentials are structured or interpreted, and the data they hold
- ▶ Coerced participation and/or data sharing
- ▶ Continued and/or increased marginalization of people whose experiences don't map cleanly onto LWR and VDC formats, frameworks, or expectations

Moreover, if proprietary vendors, private funders, or government mandates continue to dominate the record and credentialing ecosystem without deep and intentional consultation and codesign with learners and workers, individuals may lose even more control over their learning and work data and information, which is the exact opposite of what many in the ecosystem are promising as an outcome or impact.

CONCLUSION

As we shared in the opening chapter, imagine a verifiable digital credential issued to a formerly incarcerated individual who completed a skilled trades program to support their transition or reentry into the workforce. That credential could help them prove their abilities and get a good job and quickly, or it could mark them in a way and permanetly that results in employers excluding them without context or human understanding of their experience and abilities.

What changes the outcome? The policies, design decisions, and trust frameworks around how the record and credential ecosystem and associated technologies are designed, deployed, and used.

The rest of this book explores how and where harm occurs, at the level of data structures, governance, and human experience, making visible biases, discrimination, and the role of social power. The next chapter begins our conversation with one of the most common points of concern and harm: *data privacy and consent.*

FURTHER READING

1. 1EdTech. (2024). *Open Badges 3.0 specification.* https://www.imsglobal.org/spec/ob/v3p0/

2. Credential Engine & Digital Credentials Consortium. (2024). *Issuer registry research project.* https://credreg.net/announcements

3. Digital Credentials Consortium. (2020). *Building the digital credential infrastructure for the future* [White paper]. https://digitalcredentials.mit.edu/wp-content/uploads/2020/02/white-paper-building-digital-credential-infrastructure-future.pdf

4. Internet Engineering Task Force. (2024). *Selective disclosure for JWTs (SD-JWT)* [Internet-Draft]. https://datatracker.ietf.org/doc/draft-ietf-oauth-selective-disclosure-jwt/

5. Jobs for the Future. (2024). *Realizing the potential of learning and employment records as verifiable credentials: A road map for successful demonstrations.* https://www.jff.org/blog/realizing-the-potential-of-learning-and-employment-records-as-verifiable-credentials-a-road-map-for-successful-demonstrations/

6. Mozilla Foundation. (2022). *Mozilla Foundation.* https://foundation.mozilla.org/

7. T3 Innovation Network. (2024). *LER resource library.* Learn & Work Ecosystem Library. https://learnworkecosystemlibrary.com/initiatives/ler-resource-library-t3-innovation-network/

8. World Wide Web Consortium. (2025). *Verifiable credentials data model 2.0.* https://www.w3.org/TR/vc-data-model-2.0/

CHAPTER

04

DATA PRIVACY AND CONSENT

- ▶ Selective disclosure versus forced transparency
- ▶ The myth of informed consent in digital systems
- ▶ Risks of over-disclosure in hiring, education, and services
- ▶ Real-world example: healthcare, justice-involved individuals

Can technologies designed to enhance privacy actually erode it? When credentials become verifiable, permanent, and portable, do individuals gain control, or lose it?

This chapter investigates these questions by examining how learning and work records function within digital credential ecosystems. It explores the disconnect between the technical architecture of selective disclosure and the social reality of coercive data requests, the illusion of consent in contexts where refusal carries consequences, and the permanence of digital records that individuals cannot truly delete or control. Through cases ranging from COVID-19 health credentials to employment verification systems, it reveals how privacy violations can be embedded in the very design of 'privacy-enhancing' technologies.

This chapter explores the ways in which LWRs issued as VDCs can lead to privacy violations and weaken meaningful consent.

THE MYTH OF VOLUNTARY CONSENT

Consent in digital ecosystems is rarely genuinely voluntary. While individuals may be asked to 'consent' to sharing their credentials, this often occurs under conditions of structural coercion, needing a job, accessing public benefits, or participating in training programs. When the alternative to consent is exclusion from essential opportunities, the choice becomes illusory. This coerced consent is not consent at all.

Consider the contemporary job seeker navigating online application systems. To be considered for a single position, candidates typically must create accounts on platforms such as LinkedIn, Indeed, or company-specific applicant tracking systems (ATS), each demanding credential disclosure. A candidate applying to multiple positions shares their complete educational history, employment records, certifications, and references repeatedly across different platforms and organizations.

Each submission creates a digital trail over which applicants have minimal control. According to Jobscan's 2024 research, 98.4% of Fortune 500 companies use applicant tracking systems to manage recruitment (Jobscan, 2024). While GDPR-compliant systems in Europe mandate data retention limits (typically 6-12 months for unsuccessful candidates), many systems, particularly those operating under less stringent regulations, retain candidate data indefinitely unless deletion is specifically requested.

The asymmetry is stark: employers gain comprehensive information, while applicants lack visibility into data handling, retention periods, or secondary uses. Research indicates that ATS platforms routinely aggregate and analyze applicant data, though transparency about these practices remains limited.

Furthermore, most systems demand comprehensive disclosure rather than selective sharing. A candidate cannot easily share only relevant experiences or qualifications while withholding potentially prejudicial information such as employment gaps, over-qualification, or credential metadata that may reveal age or socioeconomic background.

The choice is typically binary: disclose everything or forfeit consideration.

Even more troubling, consent in these contexts is neither time-limited nor revocable in practice. Applicants who are not hired generally cannot request data deletion outside GDPR-protected regions. Those who secure positions elsewhere cannot revoke previous employers' ongoing access to their records, credentials and information. The 'consent' granted during application becomes permanent permission.

Verifiable digital credentials, despite promises of selective disclosure, often replicate these problems. When employers request LWRs issued as VDCs, they may demand comprehensive credential bundles as a condition of consideration. When platforms intermediate VDC exchanges, they create additional points of data collection.

The Consentful Tech Project defines meaningful consent using the F.R.I.E.S. framework: consent must be Freely given, Reversible, Informed, Enthusiastic, and Specific. By this standard, consent in most digital record and credential ecosystems fails on multiple criteria.

1. Applicants are not fully informed about downstream data uses.

2. Their consent is not specific to particular purposes or time frames.

3. And access and consent is certainly not easily revocable once granted.

True voluntary consent requires genuine alternatives, meaningful control, and the real possibility of refusal without penalty. In contexts where digital records and credentials, especially verifiable ones, gate access to work and learning opportunities, these conditions are rarely met.

OVER-DISCLOSURE IN PRACTICE

Verifiable digital credentials contain both explicit claims (such as degree earned or employment status) and meta-data that can reveal sensitive information beyond what the holder intends to share. Credential metadata typically includes issuer identifiers, issuance dates, credential types, validity periods, and cryptographic proof mechanisms, as well as rich-data used to provide or include evidence.

When credentials are shared, this metadata can enable inferences about protected characteristics:

▸ **Educational and institutional affiliations** may correlate with socioeconomic status, geographic location, or demographic characteristics

▸ **Program-specific credentials** (such as those issued by rehabilitation programs, disability support services, or immigration assistance organizations) can inadvertently disclose sensitive status information

▸ **Temporal metadata** (issuance dates, validity periods) may reveal gaps in employment or education that holders prefer to contextualize themselves

▸ **Issuer identifiers** that reference specialized institutions can signal disability status, immigration circumstances, or justice system involvement

While the W3C Verifiable Credentials Data Model v2.0 emphasizes selective disclosure as a privacy-enhancing feature, research on accessibility-related credentials notes that 'attestation of disability status obviously does involve this assertion' and carries 'very considerable privacy implications'. Even when holders can choose which records or credentials to present, the metadata embedded within those credentials may reveal more than the explicit claims alone.

These privacy concerns are particularly acute for people from marginalized groups and communities. As one analysis of self-sovereign identity observes, records and credentials designed to facilitate access to benefits or services may simultaneously create 'permanent digital records' that follow individuals across contexts.

Without support for technologies like *Selective Disclosure JWTs (SD-JWTs)* or *Zero-Knowledge Proofs (ZKPs)*, verifiable digital credentials can only be presented in full, making them unusable in sensitive contexts. People may avoid using them altogether, or share them reluctantly, weakening the entire trust framework.

DATA PERMANENCE AND THE ILLUSION OF CONTROL

Once issued, verifiable digital credentials exist as cryptographically signed data objects that can be copied, cached, logged, or stored by any party who receives them. While credential holders may control when and where to initially present a credential, they cannot control what happens to that data afterward.

The W3C Verifiable Credentials specification acknowledges this limitation, noting that 'verifiable presentations are usually short-lived' and 'not meant to be stored for a longer period,' yet it provides no technical mechanism to enforce this recommendation. Once a holder shares a credential with a verifier, the holder has no technical means to:

▸ Prevent the verifier from storing the record or credential data indefinitely

▸ Require deletion of the record or credential data after verification

- Detect whether the record or credential data has been shared with third parties
- Revoke access to previously disclosed information

This represents a fundamental mismatch between the promise of 'human-individual control' and the technical reality of data permanence. As research on COVID-19 health and vaccination credentials notes, 'a verifiable credential document can be stored on any device. Its traceability is not linked or attached to a person or device'.

The concept of 'revocation' in verifiable credential systems typically refers to an issuer's ability to invalidate a credential, yet not a holder's ability to retract previously shared information. Holders may be able to stop presenting a credential going forward, but they cannot undo past disclosures or force verifiers to delete stored data.

This creates particular risks in employment, education, and government benefit contexts where:

- **Rejected applications** may result in permanent credential records held by organizations that denied the applicant
- **Regulatory compliance** may incentivize indefinite data retention despite holder preferences
- **Aggregation across verifiers** can create shadow profiles the holder never authorized
- **Lack of data protection enforcement** in many jurisdictions leaves holders with no meaningful recourse

True data self-sovereignty would require not just cryptographic control over credential presentation, but also legal and technical mechanisms ensuring holders can limit data retention, require deletion, and obtain redress when these rights are violated.

Without such protections, 'holder control' remains illusory, limited to the initial moment of disclosure rather than extending to the full lifecycle of the data and its use.

CASE EXAMPLE: DIGITAL HEALTH CREDENTIALS DURING COVID-19

During the COVID-19 pandemic, digital health credentials, such as vaccine card, passports and vaccination verification apps, were introduced globally as public health tools. Beginning in early 2021, countries including Israel, France, Italy, and several U.S. states implemented systems requiring proof of vaccination for access to workplaces, restaurants, entertainment venues, and travel. Many of these systems were deployed rapidly with limited transparency about data practices and minimal learner or worker control over information sharing.

Workers in public-facing sectors, including healthcare, education, hospitality, and retail, were often required to share vaccination credentials with employers, customers, or regulators. While some systems, like New York's Excelsior Pass, claimed not to track a persons location, privacy experts noted that the separate verification apps used by businesses to scan QR codes could potentially log individual movements and retain this data. People frequently had no visibility into how long verification apps retained their data, whether it was shared with third parties, or if it could be used for purposes beyond immediate health verification, such as marketing analytics, law enforcement requests, or immigration enforcement.

The lack of uniform technical standards meant that privacy protections varied dramatically across implementations. Some systems, like Illinois's COVID-19 portal, required people to provide Social Security numbers and temporarily unfreeze their credit reports through Experian for identity verification, creating both privacy risks and barriers to access.

Canada's Privacy Commissioner explicitly warned that 'active tracking or logging of an individual's activities through a vaccine passport, whether by app developers, government, or any third party, should not be permitted,' yet enforcement mechanisms for such principles remained largely absent. Research documented that many existing vaccine passport systems had 'privacy vulnerabilities' with no formal security guarantees, leaving individuals unable to prevent indefinite data retention, detect unauthorized sharing, or obtain redress when their information was misused.

People from marginalized communities, especially those with histories of immigration, criminalization, or institutionalization, are most vulnerable to privacy harms. What seems like a neutral credential may become a source of risk, profiling, or exclusion. When people fear sharing their credentials, they are less likely to benefit from systems that were supposed to empower them. Consent becomes a formality rather than a safeguard.

ENSURING PRIVACY AND CONSENT DRIVEN GOVERNANCE

Building these digital and verifiable learning and work record and credentials ecosystems is not only a technical undertaking but also a legal and ethical one. Because learning and work records are used to aggregate sensitive information about a person's learning, work, identity, and life, they must be designed with robust privacy protections and consent-led governance frameworks.

This responsibility grows more complex in a world where learners and workers are increasingly mobile, and records often cross local, national, and system borders.

In the United States, the privacy landscape is fragmented but guided by several key laws. At the federal level, the Family Educational Rights and Privacy Act (FERPA) restricts the disclosure of student education records without explicit consent. Beyond federal law, states like California, Colorado, and Virginia have enacted comprehensive consumer privacy acts that extend rights for individuals to access, correct, and control their personal information. Compliance with these laws means that any institution or employer issuing credentials must ensure that records can only be accessed with proper authorization and that learners retain meaningful control over their data.

Internationally, the regulatory environment is even more diverse. The European Union's General Data Protection Regulation (GDPR) sets one of the most rigorous standards in the world, requiring data minimization, explicit consent, the right to erasure, and strong accountability measures for institutions handling personal data. Countries such as Canada and Australia have enacted national privacy frameworks with similar principles, while others are actively drafting legislation in response to the growing digital economy.

For any learnng and work system intended for international use, these differences cannot be ignored. A record designed in the United States but shared with an employer in Germany or a university in Australia must meet the requirements of all applicable jurisdictions.

To navigate these complexities, we must adopt a privacy-first and consent-always approach that transcends borders. Individuals should be able to give, revoke, and audit consent in clear and accessible ways, regardless of jurisdiction. Minimal disclosure, enabled by technologies like selective disclosure and zero-knowledge proofs, ensures that only the necessary facts are shared, such as proving eligibility for a job without exposing the full record of study.

Trust frameworks must be established so that issuers, verifiers, and holders of record and credentials systems can operate within interoperable and transparent systems that respect all international standards.

Strong governance is what ultimately binds these commitments together. Institutions issuing learning and work records and credentials must be accountable not only to their local legal frameworks but also to the global trust networks their learners and workers engage with. The challenge is balancing innovation with regulation.

Considerations may include:

1. **Implement Legal and Data Governance and Accountability Frameworks from the Start.** Define roles and responsibilities for data ownership and stewardship, and set up a governance body or oversight mechanism that monitors compliance.

2. **Lead with Privacy by Design.** Instead of treating privacy as an afterthought, integrate it into the architecture and workflows from day one. That means limiting data collection to what is necessary, building for minimal disclosure, and adopting principles like data minimization and consent management at the design stage.

3. **Map Applicable Laws and Standards.** Identify the regulatory frameworks that apply to the project, both within the U.S. (FERPA, HIPAA, state-level privacy acts) and internationally (GDPR, Canada's PIPEDA, Australia's Privacy Act).

4. **Establish Clear Consent and Control Mechanisms.** Ensure individuals can grant, revoke, and audit consent over the use of their data. Consent should be transparent, easy to understand, and actionable. Build tools for selective disclosure, so people can share only what's necessary in a given context.

5. **Prioritize Transparency and Trust with Ecosystem Partners.** Publish clear policies and data management plans about how data is collected, stored, used, and shared. Engage learners, employers, and institutions in the governance process to build trust. Transparency not only fulfills regulatory requirements but also strengthens adoption by demonstrating accountability.

CONCLUSION

Learning and work records issued as verifiable digital credentials or open badge standards, were originally designed to give people more control over their data and information.Yet, in practice they often reproduce the same asymmetries of power they claim to solve.

When they can only be shared in full, when consent is required under coercive conditions, and when data persists beyond the holder's control, *privacy becomes a promise* rather than a reality. For people from marginalized communities, these risks are amplified, turning tools meant to empower into tools that profile, exclude, or surveil.

True privacy in learning and work records cannot be achieved through cryptography or technology alone. It requires systems designed around minimal disclosure, revocable consent, clear accountability, and privacy-by-design from the outset. It requires governance structures that are transparent, participatory, and capable of enforcing these principles across borders and jurisdictions. It requires a shift in culture, from treating consent as a checkbox and data as an organization assets, to recognizing learning and work data as a personal assets, and consent as a continual, mutual negotiation of trust.

If LWRs and VDCs are to fulfill their promise, they must be built as public-interest infrastructure, privacy-first and consent-always, not merely as digital replicas of old record keeping systems. By centering privacy, selective disclosure, and consent-led governance, we can ensure that people, not systems, retain power over their own records, and that VDCs become instruments of equity rather than instruments of control.

FURTHER READING

1. Ada Lovelace Institute. (2021). *Checkpoints for vaccine passports.* https://www.adalovelaceinstitute.org/report/checkpoints-for-vaccine-passports/

2. And Also Too. (2018). *Building consentful tech* [Zine]. Consentful Tech Project. http://www.consentfultech.io/

3. ATZ CRM. (2025). GDPR recruitment: How to handle candidate data in 2025. https://atzcrm.com/blog/gdpr-recruitment-candidate-data/

4. Black, J. S., van Esch, P., & Arli, D. (2022). AI-enabled recruiting algorithms and the prospective employee experience: Contemporary socio-technical challenges and opportunities. *Personnel Review, 52*(9), 2369-2391. https://doi.org/10.1108/PR-10-2021-0751

5. Brennan Center for Justice. (2021). Evaluating the privacy and equity concerns posed by digital vaccine credentials. https://www.brennancenter.org/our-work/analysis-opinion/evaluating-privacy-and-equity-concerns-posed-digital-vaccine-credentials

6. Dencik, L., & Sanchez-Monedero, J. (2022). Data justice. *Internet Policy Review, 11*(1), 1–16. https://doi.org/10.14763/2022.1.1615

7. European Union Agency for Fundamental Rights. (2020). *Getting the future right: Artificial intelligence and fundamental rights.* https://fra.europa.eu/en/publication/2020/artificial-intelligence-and-fundamental-rights

8. Frontiers in Blockchain. (2024). Challenges of user data privacy in self-sovereign identity verifiable credentials for autonomous building access during the COVID-19 pandemic. *Frontiers in Blockchain.* https://www.frontiersin.org/journals/blockchain/articles/10.3389/fbloc.2024.1374655/full

9. Hireserve. (2024). Robust data protection (GDPR) solutions with Hireserve ATS. https://hireserve.com/data-protection/

10. iSmartRecruit. (2024). How applicant tracking systems ensure GDPR compliance? https://www.ismartrecruit.com/blog-gdpr-compliance-ats-recruiting-software

11. Jobscan. (2024). 2023 applicant tracking system (ATS) usage report: Key shifts and strategies for job seekers. https://www.jobscan.co/blog/fortune-500-use-applicant-tracking-systems/

12. McGregor, L., Murray, D., & Ng, V. (2019). International human rights law as a framework for algorithmic accountability. *International and Comparative Law Quarterly, 68*(2), 309–343. https://doi.org/10.1017/S0020589319000046

13. Office of the Privacy Commissioner of Canada. (2021, May 19). Privacy and COVID-19 vaccine passports. https://www.priv.gc.ca/en/opc-news/speeches-and-statements/2021/s-d_20210519/

14. Raghavan, M., Barocas, S., Kleinberg, J., & Levy, K. (2020). Mitigating bias in algorithmic hiring: Evaluating claims and practices. In *Proceedings of the 2020 Conference on Fairness, Accountability, and Transparency* (pp. 469-481). https://doi.org/10.1145/3351095.3372828

15. Solove, D. J. (2006). A taxonomy of privacy. *University of Pennsylvania Law Review, 154*(3), 477–564. https://doi.org/10.2307/40041279

16. Sporny, M., Guy, A., Sabadello, M., & Reed, D. (2022). *Decentralized identifiers (DIDs) v1.0: Core architecture, data model, and representations.* W3C Recommendation. https://www.w3.org/TR/did-core/

17. SpringerLink. (2023). Privacy-preserving digital vaccine passport. In *Advances in Information and Computer Security* (pp. 137-156). https://link.springer.com/chapter/10.1007/978-981-99-7563-1_7

18. White, J. J. (2020, January 8). Some potential accessibility-related verifiable credentials use cases. W3C public-rqtf mailing list. https://lists.w3.org/Archives/Public/public-rqtf/2020Jan/0014.html

19. World Wide Web Consortium. (2025). *Verifiable credentials data model v2.0.* W3C Recommendation. https://www.w3.org/TR/vc-data-model/

05

THE INVISIBLE BIASES
IN DIGITAL RECORDS
AND CREDENTIALS

- ▶ How credential design encodes power and exclusion
- ▶ Algorithmic sorting and data-driven discrimination
- ▶ Equity wash: issuing credentials ≠ equitable recognition
- ▶ Example: refugee job seekers, marginalized learners

Bias is often viewed as an interpersonal problem, a matter of prejudice or misjudgment between people. Yet in digital systems, bias is in the design of the data and the system. It becomes systemic, codified into algorithms, taxonomies, metadata, schemas, and governance frameworks.

Learning and Work Records (LWRs) issued as Verifiable Digital Credentials (VDCs) carry the promise of merit-based hiring and inclusion. However, they can just as easily reinforce the same historical and structural inequities that have long excluded marginalized communities from education and employment opportunities.

This chapter explores how LWRs issued as VDCs can reflect and amplify bias across design, implementation, and interpretation, and what that means for workers and learners from underrepresented and marginalized groups.

TECHNOLOGICAL NEUTRALITY IS A MYTH

The perception that technology is neutral persists, even in the face of mounting evidence to the contrary. From facial recognition software that fails to identify Black faces to credit-scoring algorithms that discriminate based on cultural context, zip or postcode, technology systems often reproduce the inequities of the world around them.

In digital record or credentialing systems, this neutrality myth appears when technologists assume that all credentials, all issuers, and all forms of knowledge are treated equally in a digital environment. Yet they're not.

A 2020 study by the *AI Now Institute* emphasized that AI systems often replicate institutionalized biases, especially when trained on skewed or incomplete data sources. The same is true of digital record and credentialing. Systems that prioritize formal degrees, exclude grassroots issuers, or reward specific types of experiences such as internships over community organizing, embeding inequity into the record and credentialing landscape.

WHAT COUNTS AS A SKILL AND WHO DECIDES?

A taxonomy is at the heart of every learning and work record and credential system (be it digital, verifiable or not). A taxonomy is a classification of skills, competencies, behaviors, and knowledge. Yet the creation of these taxonomies is deeply political. What is defined as a skill, and what is not, is often based on dominant cultural, racial, and economic perspectives. Consider this:

▸ Is navigating the child welfare system for your family a skill?

▸ Is organizing tenants for a rent strike a skill?

▸ Is surviving incarceration a skill?

These capacities involve negotiation, communication, resilience, leadership, systems thinking, and yet they are rarely included in standard skill frameworks like the Occupational Information Network (O*NET) or the European Skills, Competences, Qualifications and Occupations (ESCO) standard skills framework, which power many LWR tools, platforms, and systems.

This exclusion leads to a phenomenon Ruha Benjamin calls 'informed exclusion,' where systems know what they're leaving out and proceed anyway.

*'The power to define what is valid knowledge is a political act.' Ruha Benjamin, *Race After Technology* (2019)*

ALGORITHMIC SORTING AND RANKING

As learning and work records issued as verifiable digital credentials, become machine-readable, they are increasingly used in algorithmic systems that filter, rank, and score candidates. These systems often rely on credential data and metadata, such as issuer name, credential type, and date, to determine how 'qualified' someone is. This creates several bias risks (Figure 5):

Figure 5. Record and Credential Bias and Risks

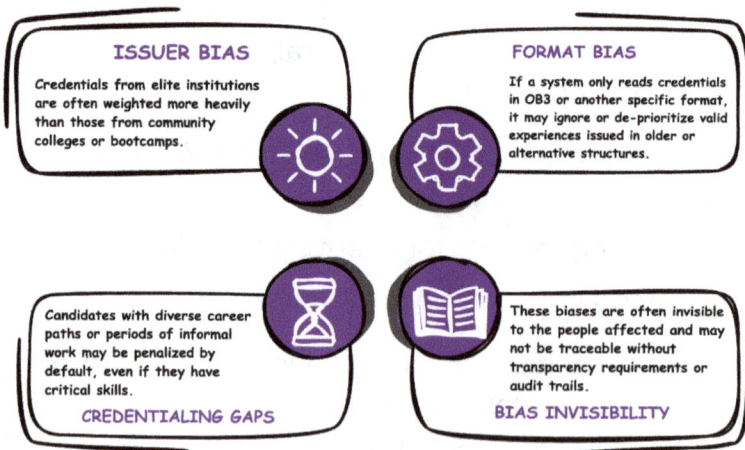

ISSUER BIAS
Credentials from elite institutions are often weighted more heavily than those from community colleges or bootcamps.

FORMAT BIAS
If a system only reads credentials in OB3 or another specific format, it may ignore or de-prioritize valid experiences issued in older or alternative structures.

CREDENTIALING GAPS
Candidates with diverse career paths or periods of informal work may be penalized by default, even if they have critical skills.

BIAS INVISIBILITY
These biases are often invisible to the people affected and may not be traceable without transparency requirements or audit trails.

© LWYL Studio

▶ **Issuer bias:** Credentials from elite institutions are often weighted more heavily than those from community colleges or bootcamps.

- **Format bias:** If a system only reads credentials in OB3 or another specific format, it may ignore or de-prioritize valid experiences issued in older or alternative structures.

- **Credentialing gaps:** Candidates with diverse career paths or periods of informal work may be penalized by default, even if they have critical skills.

- **Bias invisibility:** These biases are often invisible to the people affected and may not be traceable without transparency requirements or audit trails.

EXCLUSION THROUGH LANGUAGE AND FRAMING

Many learning and work records and verifiable digital credentials use standardized frameworks that are:

- Written in academic or technical English

- Designed around Western business norms

- Based on assumptions of continuous formal employment

For immigrants, refugees, or non-English speakers, the language of these credentials may be challenging to navigate or irrelevant to their lived experience. In a 2021 presentation from the *Open Recognition Alliance*, community-led credentialing efforts in Indigenous, African, and Latin American contexts emphasized the need for *vernacular credentials,* locally understood, culturally grounded, and linguistically relevant.

Yet most credentialing vendors do not support multilingual templates or culturally responsive schemas.

GATEKEEPING THROUGH ISSUER REGISTRIES

Record and credentialing systems that use issuer allowlists or registries (e.g., only allowing universities, government agencies, or Fortune 500 companies to issue credentials) may also exclude grassroots organizations, small businesses, cooperatives, soloentrepreneurs or freelances, unions, and mutual aid networks. This gatekeeping:

- ▶ Undermines the credibility of community-led training
- ▶ Reduces recognition for informal or 'underground' skills
- ▶ Privileges institutional validation over lived experience
- ▶ Excludes individuals issuing through self-assertions and endorsements.

A 2022 study by the *Credential As You Go* initiative found that employers disproportionately value credentials from well-known entities even when they are functionally identical to those issued by local programs.

CASE EXAMPLE: UNDERRECOGNITION OF HBCUs AND TRIBAL COLLEGES

Credential Engine's 2023 Equity Advisory Council Report on credential transparency highlighted systemic gaps in how credentials from minority-serving institutions are represented in national registries and employer-facing platforms. While the report did not specifically single out Historically Black Colleges and Universities (HBCUs) and Tribal Colleges and Univiersities (TCU) by name, it emphasized that populations experiencing racial minoritization face significant barriers in credential recognition, a pattern that disproportionately affects graduates from institutions serving these communities.

These institutions serve critical roles in advancing education access for Black and Indigenous students. HBCUs comprise only 3% of colleges and universities yet enroll 16% of all Black students in higher education, producing 22% of bachelor's degrees granted to African Americans. Similarly, the 35-37 Tribal Colleges and Universities serve approximately 28,000 students annually, with over 80% from reservation communities, yet these institutions remain "invisible" to much of the public and are frequently perceived as lower quality despite full accreditation.

Yet when employer-facing platforms and algorithmic hiring systems process credentials from these institutions, multiple forms of bias emerge. Research on algorithmic hiring has documented how AI-driven resume screening tools can disadvantage candidates from institutions outside the traditional 'elite' university networks. As one NPR report noted, if recruiters' social networks lack diversity and don't include anyone who attended an HBCU or Latino-serving institution, those candidates may never even enter the applicant pool. Hiring managers have acknowledged that 'tiering of school partnerships', explicitly ranking some institutions as more prestigious than others, creates divisions and perpetuates perceptions that some institutions are inherently better, regardless of actual educational quality or graduate outcomes.

The result? Graduates with the same competencies as their peers from predominantly white institutions (PWIs) face systematic disadvantages. Despite evidence that Black HBCU graduates report better career outcomes than Black graduates from PWIs, including higher lifetime earnings, higher-status occupations, and greater job satisfaction, they remain significantly underrepresented in high-paying fields like finance (7.4%) and technology (6.2%). TCU graduates face similar challenges, compounded by geographic isolation, limited employer awareness of these institutions, and historical trauma that makes trust-building with mainstream employers particularly complex.

These patterns extend beyond individual bias into the architecture of digital hiring systems themselves. When algorithmic systems are trained on historical hiring data that reflects past discrimination, they encode and amplify institutional prestige bias. Credentials from HBCUs and TCUs may be algorithmically filtered out, deprioritized in search results, or categorized as 'other' in ways that reduce graduate visibility to employers, even when those graduates possess identical or superior qualifications. The automation of credential evaluation, rather than eliminating bias, risks making it invisible and more difficult to challenge.

Addressing this requires more than adding HBCU and TCU credentials to databases. It demands that credential registries actively audit for institutional bias, that algorithmic hiring tools be tested for disparate impact across institution types, and that employers commit to building authentic, sustained relationships with these institutions rather than one-time recruiting visits. As Credential Engine's equity framework emphasizes, transparency alone is insufficient, equity requires intentional design choices that counteract, rather than replicate, historical patterns of exclusion.

BIAS BY OMISSION: INFORMAL AND JUSTICE-IMPACTED WORKERS

Millions of workers gain skills outside formal education record or credentialing channels, including care workers, gig economy participants, formerly incarcerated individuals, immigrants, and people working in what some people call informal economies. Yet when digital record and credential systems fail to recognize these experiential pathways, they effectively render these individuals invisible to employers and algorithmic hiring platforms that depend on standardized record and credential data.

The scope of this exclusion is staggering. *The Rework America Alliance (2021)* found that approximately 60% of

U.S. workers, representing over 70 million people, do not hold a bachelor's degree, yet they have built substantial capabilities through work experience and non-formal ways of learning and education. Despite growing rhetoric around 'skills-based and skills-first hiring,' most applicant tracking systems and AI-powered recruitment platforms continue to use degree attainment as a primary filter or proxy for competence. A *Harvard Business School* study found that for middle-skills jobs, 65% of job postings for executive assistants required a college degree, yet only 19% of current employees in those roles actually held one, a phenomenon researchers termed 'degree inflation' or 'upcredentialing.'

This creates what appears to be a 'skills gap' yet it is more accurately described as a *recognition gap*, not a deficit in capability, but a failure of record and credentialing infrastructure to see and validate what people already know and do. When algorithms sort candidates based on degree completion, work history at recognized employers, or credentials listed in standardized registries, they systematically exclude:

▶ **Care workers and domestic workers**, whose skills in elder care, disability support, and household management are rarely formalized into portable credentials despite their economic and social value

▶ **Gig economy and platform workers**, who develop project management, customer service, and technical skills through Uber, TaskRabbit, or freelance work that algorithms may not categorize as "real" employment

▶ **Justice-impacted individuals**, who face both the stigma of criminal records in algorithmic background checks and the erasure of any skills gained through prison education, vocational training, or work-release programs

▶ **Immigrant workers**, whose foreign records and credentials are often unrecognized or undervalued by U.S.

credentialing systems, and whose language skills and cross-cultural competencies go unacknowledged

- **Workers in informal economies**, including those in cash-based, family businesses, or community-based work that leaves no digital trace for algorithms to parse

The exclusion is compounded by the emerging design of digital record and credential verification systems themselves. Digital badges, blockchain credentials, and competency-based records all require someone with institutional authority to award and issue them. Workers who learn through lived experience such a formerly incarcerated person who becomes a peer counselor, a grandmother who provides elder care, a community organizer who masters conflict resolution, have no institution to validate their expertise. As Virginia Eubanks documented in *Automating Inequality* (2018), automated systems designed to allocate resources and opportunities most heavily surveil and punish those who are already economically marginalized, creating a 'digital poorhouse' that mirrors historical patterns of discrimination.

Furthermore, when justice-impacted individuals do acquire credentials during incarceration, algorithmic systems may devalue or ignore them. A study on algorithmic bias in hiring found that when applicant tracking systems flag individuals with employment gaps or non-traditional work histories, formerly incarcerated individuals, even those with certifications earned through prison education programs, are disproportionately filtered out before human review.

Addressing bias by omission requires more than adding 'alternative or non-degree credentials' to existing registries. It demands:

- **Recognition ecosystems** that allow peer validation, community endorsement, and self-attestation alongside institutional credentialing

- **Algorithmic audits** that test for disparate impact not just by race and gender, but by educational pathway and work history type
- **Policy interventions** such as ban-the-box policies for algorithmic hiring, requirements for human review of filtered candidates, and portable learning and work records (LWRs) that workers control
- **Equity-centered design** that asks not 'How can we credential informal workers?' but 'How might credentialing systems adapt to recognize the full spectrum of human capability?'

Without these interventions, digital record and credential systems risk becoming another mechanism through which structural inequality is automated, amplified, and made invisible, all while appearing neutral, efficient, and fair.

TOWARD BIAS-AWARE RECORD AND CREDENTIAL DESIGN

What can be done? Ethical learning and work records and verifiable digital credential systems designed must implement the following actions (Figure 6) to mitigate bias.

Figure 6. Mitigating Bias in Credential Design

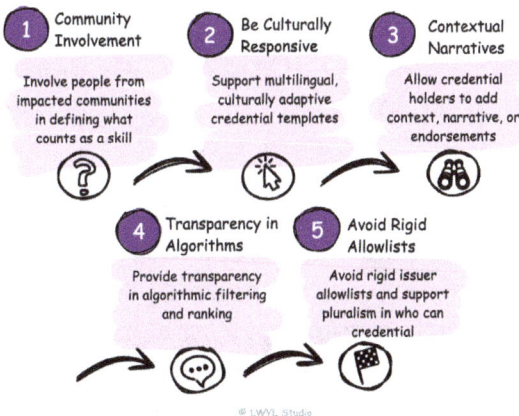

1 Community Involvement
Involve people from impacted communities in defining what counts as a skill

2 Be Culturally Responsive
Support multilingual, culturally adaptive credential templates

3 Contextual Narratives
Allow credential holders to add context, narrative, or endorsements

4 Transparency in Algorithms
Provide transparency in algorithmic filtering and ranking

5 Avoid Rigid Allowlists
Avoid rigid issuer allowlists and support pluralism in who can credential

© LWYL Studio

- ▶ **Involve** people from impacted communities in defining what counts as a skill
- ▶ **Support** multilingual, culturally adaptive credential templates
- ▶ **Allow** credential holders to add context, narrative, or endorsements
- ▶ **Provide** transparency in algorithmic filtering and ranking
- ▶ **Avoid** rigid issuer whitelists and support pluralism in who can credential

Without these changes, issuing learning and work records issued as verifiable digital credentials risks becoming just another layer in a long history of data-driven discrimination.

CONCLUSION

Verifiable digital credentials may appear neutral on the surface, but every aspect of their design, what counts as a skill, who gets to issue them, how algorithms filter them, is shaped by choices that carry political and cultural weight.

As this chapter has shown, Learning and Work Records (LWRs) and Verifiable Credentials (VDCs) can easily reproduce the very exclusions they claim to disrupt. From issuer bias to inaccessible language and rigid taxonomies, people from marginalized communities in particular face structural barriers embedded in systems meant to increase access.

Bias in verifiable digital credentials is not simply a technical flaw; it is a reflection of power, of who gets to decide whose knowledge is valid, whose skills are visible, and whose labor is worth recording. Without deliberate design for pluralism, accessibility, and community-led validation, these systems risk deepening the invisibility of informal workers, justice-impacted individuals, and historically ex-cluded institutions like HBCUs and Tribal Colleges.

Designing for equity requires more than issuing more credentials, it demands questioning the foundations of how recognition is granted and to whom. Ethical credentialing must move beyond technical interoperability to social and cultural responsiveness. By embedding bias mitigation strategies, transparency, and participatory design, we can begin to build systems that truly recognize the full spec-trum of human experience, not just what fits into a schema.

1. AI Now Institute. (2019). Discriminating systems: Gender, race, and power in AI. https://ainowinstitute.org/discriminatingsystems

2. Baker, R. S., & Hawn, A. (2022). Algorithmic bias in education. *International Journal of Artificial Intelligence in Education, 32*(4), 1052–1092. https://doi.org/10.1007/s40593-021-00285-9

3. Benjamin, R. (2019). *Race after technology: Abolitionist tools for the new Jim code.* Polity Press.

4. Credential As You Go. (2022). *Employer value and equity in credentialing.* https://credentialasyougo.org/resources/

5. Credential Engine. (2023). *Equity advisory council report and recommendations.* https://credentialengine.org/credential-transparency/equity/

6. Diverse Issues in Higher Education. (2024). Tribal colleges and universities seek greater recognition and funding. https://www.diverseeducation.com/institutions/tribal-colleges/article/15305956/

7. D'Ignazio, C., & Klein, L. F. (2020). *Data feminism.* MIT Press. https://data-feminism.mitpress.mit.edu/

8. Eubanks, V. (2018). *Automating inequality: How high-tech tools profile, police, and punish the poor.* St. Martin's Press.

9. National Association of Colleges and Employers (NACE). (2021). Challenges, solutions for college recruiting at HBCUs. https://www.naceweb.org/diversity-equity-and-inclusion/best-practices/challenges-solutions-for-college-recruiting-at-hbcus/

10. Noble, S. U. (2018). *Algorithms of oppression: How search engines reinforce racism.* NYU Press.

11. Open Recognition Alliance. (2021). *Open recognition practices: Recognition in context.* https://openrecognition.org

12. Raghavan, M., et al. (2020). Fairness and bias in algorithmic hiring: A multidisciplinary survey. *ACM Transactions on Intelligent Systems and Technology.*

13. Rework America Alliance. (2021). *The new talent landscape.* Markle Foundation. https://www.markle.org/alliance-resources/

14. Veris Insights. (2024). The current state of HBCU recruiting strategy. https://verisinsights.com/blogs/hbcu-recruiting-strategy/

06

SURVEILLANCE, TRACKING, AND CONTROL

CHAPTER INSIGHTS

▶ Linking credentials to digital ID and location tracking

▶ Employer surveillance and behavioral monitoring

▶ Government misuse and chilling effects on communities

▶ Cross-border implications and global data colonialism

The rise of Verifiable Digital Credentials (VDCs), and the issuance of learning and works records as VDCs has been framed as a technological advance and system innovation toward greater transparency, portability, and trust in learning and employment systems. Yet, for many individuals, especially those already subject to institutional surveillance, these same systems can introduce new monitoring, profiling, and control mechanisms.

In this chapter, we explore how learning and work records and credentials issued as VDCs can function as tools of surveillance and how their design, build, and deployment can threaten civil liberties, privacy rights, and individual agency.

SURVEILLANCE IS A FEATURE, NOT A BUG

Surveillance is the systematic collection, analysis, and use of data to observe or influence behavior. In the context of issuing learning and work records as VDCs, surveillance can be *passive* such as credential logs or accessing other record or credential metadata, or *active* such as linking credentials to location, biometric data, or behavioral scores.

The problem isn't just data collection and its use, it's the power asymmetries that determine who is being watched, who is watching, and for what purpose. These asymmetries are not accidental; they are embedded in the design of digital systems that serve institutional interests over individual autonomy. When credentials become both proof of achievement and instruments of monitoring, they transform from tools of opportunity into mechanisms of control.

Digital identity systems, when layered with record and credentialing platforms, risk creating persistent, portable identifiers that allow individuals to be tracked across services, employers, borders, and time. Unlike physical credentials that can be selectively presented or left at home, digital verifiable records and credentials create permanent trails of verification requests, access patterns, and cross-institutional data flows. This is especially concerning for workers in low-wage, high-turnover industries with high mobility yet weak protections, who face the greatest exposure to surveillance while having the least power to contest or refuse it.

The promise of portability becomes a double-edged sword: credentials that move with you also create records that follow you. Every verification, every job application, every border crossing generates metadata that can be aggregated, analyzed, and weaponized against the credential holder, often without their knowledge or meaningful consent.

'Data systems that follow people throughout their lives create a digital leash, one that restricts, rather than enables, freedom.'
— *Virginia Eubanks, Automating Inequality (2018)*

THE INFRASTRUCTURE OF TRACKING

Verifiable digital credentials rely on metadata, timestamps, issuer registries, wallet interactions, and verification logs. If stored or linked together or using server retrieval features, these system components can become tools for monitoring behavior. Examples include:

- ▸ **Holder identity:** Who the record or credential is linked/attched to, and the identity assets used

- ▸ **Verifier logs:** When and how often a credential is verified, and by who

- ▸ **Issuer identity:** Which institutions are linked to which individuals, and issuing what/which records and credentials

- ▸ **Wallet metadata:** Devices, IP addresses, or identifiers linked to usage

- ▸ **Credential audit trails:** Logs of where and with whom credentials are issued, shared, and verified

In theory, many of these can be anonymized or handled in a privacy-preserving and consent-driven way.

In practice, few systems currently deploy such safeguards.

GOVERNMENT-ISSUED IDS AS VDCs: CROSS-SYSTEM TRACKING

Explored in a case example in Chapter 12 is the move toward issuing government issued identity documentation such as driver's licenses, green cards, visas, and social services IDs as verifiable digital credentials, either using the mDL or the W3C VCDM. This move can open the door to cross-system surveillance and access. For example a government-issued digital ID is often used or required to:

- ▸ Purchase alcohol

- ▸ Rent an apartment

- ▸ Apply for a job

- ▸ Access public benefits

- ▸ Board a bus or train

- ▶ Enter a public building

If all these interactions generate verification logs or are tied to an individual's unique identifier, the result is a *comprehensive surveillance system*, often without the individual's full knowledge or consent as to what is being collected, by whom, and for what purpose.

In countries like China and India, digital ID systems such as Aadhaar have already demonstrated how quickly this infrastructure can be used to control access to resources or punish dissent. In the U.S., location and identity data in systems like ImmigrationOS have been used to prosecute abortion seekers, immigrants, and protestors, with increasing investment in Palantir data-driven systems.

WORKER SURVEILLANCE AND WORKFORCE CREDENTIALING

In workforce contexts, employers may use verifiable digital records and credentials to:

- ▶ Track training completion and skill acquisition
- ▶ Monitor compliance with certifications and regulations
- ▶ Assess performance against benchmarked standards
- ▶ Predict 'risk' of attrition, misconduct, or unionization

When credentials include behavioral assessments, social-emotional skill scores, or psychometric data, this information can feed into expanded surveillance practices such as:

- ▶ Real-time productivity tracking and algorithmic management
- ▶ Pre-employment screening and automated hiring decisions

- ▸ Predictive analytics for workforce planning
- ▸ Automated disciplinary actions or performance improvement plans

The permanence problem: Unlike traditional employment records that remain siloed within individual organizations, VDCs create portable surveillance profiles. Workers may carry algorithmic judgments, based on productivity metrics, behavioral scores, performance reviews, or even whistleblowing history, across jobs, industries, and borders.

In 2021, Amazon reportedly used internal performance tracking tools to monitor warehouse workers' task completion rates and issued automated warnings or terminations based on algorithmic thresholds. Workers described the system as dehumanizing and opaque, with no meaningful way to contest decisions. If such systems are extended with permanent, portable VDCs, workers could carry these 'performance scars' across their entire careers, facing discrimination based on metrics they cannot see, challenge, or correct.

The concern intensifies when record and credentials are linked to digital identity systems. A worker flagged for 'low productivity' at one Amazon warehouse could find that score trailing them to other logistics companies, retail positions, or gig platforms, creating a permanent underclass of workers marked by algorithmic judgment.

As the AI Now Institute warns, this transforms employment surveillance from a workplace issue into a lifetime sentence.

SURVEILLANCE IN EDUCATION

Schools and colleges are increasingly issuing digital badges and verifiable digital credentials tied not only to academic achievement, but also to attendance patterns, behavior incidents, participation in interventions or special education services, social-emotional learning assessments, and volunteer activities. While proponents frame these credentials as comprehensive portraits of student development, they simultaneously create detailed surveillance records that extend far beyond traditional transcripts.

These data points can be stored indefinitely and follow students from elementary school through high school, into postsecondary systems, and ultimately into job markets. Without strict controls on what can be recorded and credentialed, who can access these records, and how long they persist, digital credentials risk becoming digital dossiers that label students for life, transforming temporary challenges into permanent markers of identity.

A number of key activities and harms may include:

1. **Differential surveillance and encoded inequality:** The risks are not distributed equally. Students of color, disabled students, low-income students, and students with learning differences are disproportionately subject to disciplinary interventions, behavioral monitoring, and absenteeism tracking due to systemic biases in school discipline and special education referral practices. When these experiences are encoded into portable verifiable digital credentials, they carry forward the structural inequalities embedded in educational systems.

   Consider a student flagged for 'disruptive behavior' in middle school, a label applied disproportionately to Black and Latino boys, who later seeks to present credentials for college admission or employment. If behavioral data is included in their credential wallet, they may face automated rejections based on past disciplinary incidents, regardless of context, growth, or the discriminatory practices that led to the initial flag. The credential becomes a mechanism for perpetuating rather than remedying educational injustice

2. **The social-emotional learning trap:** Increasingly, schools are recording and credentialing 'social and emotional skills' (also referred to as soft or essential skills) like collaboration, resilience, growth mindset, and emotional regulation. While these may seem benign, they introduce subjective, culturally-biased assessments into permanent records. A student who challenges authority might be marked as lacking 'collaboration skills.' A student managing trauma might be flagged for 'emotional dysregulation.' These judgments, once credentialed, become reified as objective truth, following students into spaces where educators lack context and employers lack training to interpret them fairly.

   Moreover, the very act of measuring and credentialing social-emotional behaviors changes the student-teacher relationship, transforming every interaction into potential data for future judgment. Students learn to perform compliance rather than develop authentic skills, understanding that resistance or vulnerability may be permanently recorded.

3. **Function creep and institutional access:** Even when records and credentials are initially designed for legitimate educational purposes, say, documenting completion of a career readiness program, mission creep is inevitable. Colleges may request access to behavioral data during admissions. Employers may demand full credential histories, including disciplinary records. Immigration authorities or law enforcement may compel disclosure. Once data exists in standardized, portable formats, pressure to share it intensifies, and students, especially

those from marginalized communities, have little power to refuse.

As one privacy advocate warned, "We're building infra-structure that will follow children for the rest of their lives, and we're doing it without asking whether we should."

POLICING AND IMMIGRATION CONTROL

Increased use of digital credentials raises the possibility of police and immigration agencies accessing credential data, directly or indirectly. For example,

▶ A credentialed job application could include visa status or national origin.

▶ Police could request access to digital wallets to verify identity and activity during stops.

▶ Border agencies could use VDCs to create watchlists based on affiliations or institutions attended.

In 2022, the *Electronic Frontier Foundation (EFF)* warned that mobile driver's license pilots could expose individuals to unconstitutional searches and data collection. Without clear boundaries, the same risks apply to all VDC-based identity systems. In 2025, Dr. Kim Hamilton Duffy, Executive Director of the Digital Identity Foundation (DIF) raised concerns about digital wallet and drivers license credential features in the mDL standard which would afford issuers with the functionality to *phone home* through the activation of server retrieval features and verification logs. This means, the issuer of the verifiable digital credential will receive log files of the credential activity when ever a holder shows, displays, or has it verified.

VENDOR LOGs AND COMMERCIAL SURVEILLANCE

Even when public agencies act ethically, private data intermedieraies, or credential vendors may not. Credential wallets and verification services often generate backend logs that:

- ▶ Track human behavior
- ▶ Store credentials presented or verified
- ▶ Build analytics profiles
- ▶ Feed marketing algorithms

This creates a risk of *commercial surveillance* where individuals' educational and employment history becomes a source of profit without consent.

Reports in 2023 by the Center for Democracy and Technology (CDT) and Common Sense Media corroborated the trend of widespread, opaque data sharing in educational technology. These reports highlighted a systemic issue in the edtech industry of sharing data with third-party services for analytics or monetization, often without clear consent or transparency from learners, workers, students, or their guardians.

CHILLING EFFECTS AND SELF-CENSORSHIP

When individuals know they are being watched, or think they might be, they may self-censor. This is a well-documented psychological response to surveillance.

In the context of verifiable learning and work records:

- ▶ Workers may **avoid applying for jobs** if doing so exposes stigmatized experiences, records, or credentials
- ▶ Students may **opt out** of certain programs or services for fear of long-term tracking

- Immigrants may **avoid** digital systems altogether to reduce risk of deportation

All of this undermines the goals of inclusion, equity, and trust in record and credentialing systems, and the reveals a different and more worrying purpose for their development and use.

DESIGN FOR SURVEILLANCE RESISTANCE

To build surveillance-resistant verifiable learning and work systems, we must follow a number of key actions (Figure 7).

Figure 7. Design for Surveillance Resistance

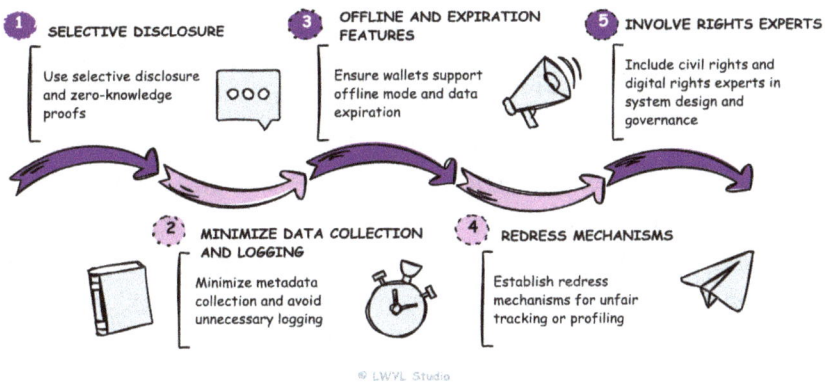

1. **SELECTIVE DISCLOSURE**
   Use selective disclosure and zero-knowledge proofs

2. **MINIMIZE DATA COLLECTION AND LOGGING**
   Minimize metadata collection and avoid unnecessary logging

3. **OFFLINE AND EXPIRATION FEATURES**
   Ensure wallets support offline mode and data expiration

4. **REDRESS MECHANISMS**
   Establish redress mechanisms for unfair tracking or profiling

5. **INVOLVE RIGHTS EXPERTS**
   Include civil rights and digital rights experts in system design and governance

© LWVL Studio

- **Use** *selective disclosure* and *zero-knowledge proofs*
- **Minimize** metadata collection and avoid unnecessary logging
- **Ensure** wallets support *offline mode* and *data expiration*
- **Establish** *redress mechanisms* for unfair tracking or profiling
- **Include** *civil rights and digital rights experts* in system design and governance
- Amongst other actions ...

Ultimately, resistance to surveillance must be a non-negotiable design principle, not a patch applied after harm is done.

CONCLUSION

Surveillance within digital record and credentialing systems is not an accidental side effect, it is a predictable outcome of how digital identity and verifiable records are being built. When learning and work records and credentials, identity data, and behavioral logs are combined into persistent, portable systems, they can quickly become mechanisms for monitoring, profiling, and controlling individuals, particularly those already subject to institutional scrutiny.

What is marketed as 'trust' or 'efficiency' can easily harden into a digital leash.

This chapter has shown that the risks of surveillance extend across contexts: from government-issued IDs to employer monitoring, from education records to cross-border tracking. Without strict privacy-by-design principles, transparent governance, and meaningful limits on data collection, verifiable credentials risk replicating some of the most harmful dynamics of digital platforms, turning people's learning and work histories into lifelong surveillance dossiers.

If our goal is to genuinely *empower people,* we require more than cryptographic security. We require active resistance to surveillance at every layer of design and policy. Selective disclosure, data minimization, expiration controls, and independent oversight are not optional, they are essential.

By embedding these safeguards from the start, we can ensure that learning and work records issued as verifiable digital credentials can really serve as tools of mobility and recognition, not instruments of tracking, surveillance, and control.

FURTHER READING

1. Ada Lovelace Institute. (2021). Participatory data stewardship: A framework for involving people in the use of data. https://www.adalovelaceinstitute.org/report/participatory-data-stewardship/

2. AI Now Institute. (2023). *Algorithmic management: Restraining workplace surveillance.* https://ainowinstitute.org/publications/algorithmic-management

3. Bigo, D. (2014). The (in)securitization practices of the three universes of EU border control: Military/Navy – border guards/police – database analysts. *Security Dialogue, 45*(3), 209-225. https://doi.org/10.1177/0967010614530459

4. Bigo, D., & Guild, E. (2005). *Controlling frontiers: Free movement into and within Europe.* Ashgate.

5. Electronic Frontier Foundation. (2021). DHS's flawed plan for mobile driver's licenses. https://www.eff.org/deeplinks/2021/07/dhss-flawed-plan-mobile-drivers-licenses

6. Eubanks, V. (2018). *Automating inequality: How high-tech tools profile, police, and punish the poor.* St. Martin's Press.

7. Lyon, D. (2018). *The culture of surveillance: Watching as a way of life.* Polity Press.

8. Privacy International. (2020). *Demanding identity systems on our terms.* https://privacyinternational.org/campaigns/demanding-identity-systems-our-terms

9. UNESCO. (2023). *Artificial intelligence and the futures of learning.* https://www.unesco.org/en/digital-education/ai-future-learning

10. Zuboff, S. (2019). The age of surveillance capitalism: The fight for a human future at the new frontier of power. PublicAffairs.

07

VENDOR LOCK-IN AND PLATFORM DEPENDENCE

CHAPTER INSIGHTS

▶ Centralized wallets in a decentralized disguise

▶ Proprietary formats and lack of portability

▶ Extractive business models disguised as 'public goods'

▶ Failed credentialing pilots and abandoned data

Learning and work record and digital credentialing systems are often framed as public infrastructure, like roads, libraries, or water systems. Yet in practice, many issuance platforms are *not public goods, community goods or even club goods*. They are built and managed by private vendors with proprietary systems, monetization strategies, and limited accountability. Although they might be informed by open standards.

This chapter explores the structural risks such as vendor lock-in, how platform dependence affects individuals and institutions, and why the realities of 'wallet, platform, or storage choice' often masks deeper power asymmetries.

THE MYTH OF DATA SELF-SOVEREIGNTY

Many digital record and credentialing platforms promise that individuals 'own' their learning records through decentralized identifiers (DIDs) and self-sovereign identity (SSI) wallets. The marketing language emphasizes control, portability, and independence: *your credentials, your data, your choice.*

The promise sounds empowering: store your credentials independently, present them selectively to employers or institutions, and maintain complete control over who accesses your information. Yet, The reality is far more constrained. When you try to actually exercise this promised sovereignty you may face four key lock-in mechanisms.

Proprietary ecosystems disguised as open standards

While many platforms claim to use 'open standards' like W3C Verifiable Credentials, their implementations often include proprietary extensions that prevent true portability. Credentials issued in one wallet therefore frequently cannot be verified by systems expecting different cryptographic proof formats. So what appears to be interoperability is often just compatibility within a vendor's partner network.

Commercial incentives trump human agency

Wallet providers are typically venture-backed companies whose business models depend on people retention and data access. The 'Free' wallets also often monetize through premium features, institutional partnerships, or data analytics. And when a wallet provider changes terms of service, pivots their business model, or shuts down, people discover they never truly 'owned' their credentials

Selective interoperability

Platforms frequently work only with credentials from approved issuers who have commercial relationships with the wallet provider. Exporting credentials in machine-readable formats may be technically possible but deliberately obscured in platform or wallet interfaces. Even when export is available, the receiving system may not accept or verify the credential format.

The technical literacy barrier

True portability would require people to understand DIDs, cryptographic keys, JSON-LD schemas, and verification methods. Most platforms offer 'people-friendly' interfaces that abstract away these details, and with them, actual control. Self-custody of private keys means self-responsibility for recovery; lose your keys, lose your credentials permanently

A Case Example

Imagine you earn a professional certificate stored in Wallet A. You want to share it with an employer who uses Verification System B. You discover:

▸ System B doesn't recognize Wallet A's credential format without a paid integration

▸ Exporting the credential produces a JSON file that System B can't verify

▸ The issuer must reissue the credential through a different platform at an additional cost

▸ Or worse: Wallet A has shut down, taking your credential verification infrastructure with it

This is not hypothetical. Early adopters of digital verifiable record and credentialing platforms have encountered all of these scenarios.

HOW VENDOR LOCK-IN WORKS

Vendor lock-in happens when people or institutions are dependent on a single or small groups of providers for functionality, storage, or access, and switching is costly or technically impossible. In digital verifiable learning and work record systems, this happens when there exists:

▸ **Platform Dependence:** Record and credentials are only readable or verifiable by the issuing vendor's software (platform dependence)

▸ **Closed or Fee-based Integrations:** The platform is closed to approved ecosystem members only, or those who pay/charge a fee to participate. Verifiers must pay to access or verify records

- **Limited Standard and Use Context Support:** Platform and wallets are proprietary and do not support open or multiple standards, use contexts, or credentials
- **Poor People-first Usability:** Present, share, and migration tools (export/import) are absent or poorly supported

EVALUATING PRODUCTS FOR LOCK-IN RISK

To evaluate a a platform or technology for *Lock-in RIsk*, LWYL Studio uses four key criteria, each of which can be evaluated for meeting either needs of a technical group of people such as developers or industry experts or people who need to use the solution to manage record and credentials for themselves, such as a learner or worker, or for an institution, such as an admin or staff member. Each group have a different role and different needs for usage.

Table 10. Lock-in Risk Evaluation Criteria

Criterion	Definition
Interoperability	How well a system's credentials and records work with external systems and solutions. Evaluates technical compatibility, standards adoption, and ability to exchange data across platforms without vendor lock-in.
Data Portability and Sharing Functionality	Evaluates the breadth and depth of features that enable individuals to share or move credential and record data. Evaluates sharing methods, export options, format flexibility, and control mechanisms.
Usability	Evaluates how easy and intuitive the solution is for individuals to manage records and credentials. Evaluates the platform or app interface design, workflow efficiency, learning curve, and practical utility.

Accessibility	Evaluates compliance with W3C Web Content Accessibility Guidelines (WCAG) and accessibility laws (ADA, Section 508). Evaluates how well the solution accommodates people with disabilities including visual, auditory, motor, and cognitive impairments.

Source: LWYL Studio (Forthcoming, 2026)

In 2025, LWYL Studio conducted a detailed *Lock-in Risk Review* of the Top Digital Record and Credential Products (forthcoming, 2026), using these criteria. A summary of the results are Table 11. See Appendix D for more information of the providers evaluated.

As shown in Table 11, only one product, SOLO, received a low risk of lock-in, scoring 30 out of 40 points across the four criteria; with 12 vendors have medium level of risk of lock-in, and the remaining 11 scoring high risk of lock-in.

Table 11. Digital Record and Credential Product Review

Vendor	Product	Risk Evalution	Score (1- 40)
The SOLO Network	SOLO	Low	30
BCdiploma	BCdiploma	Medium	27
CertifyMe	CertifyMe	Medium	26
EBSCO	EBSCOed Digital Credential Wallet	Medium	26
Accredible	Accredible	Medium	25
Gradintelligence	GradIntel	Medium	25
Greenlight Credentials	Greenlight Credentials	Medium	25
iDatafy	SmartResume	Medium	25

Instructure	Parchment + Digitary	Medium	25
Pearson	Credly Acclaim	Medium	25
SchooLinks	SchooLinks Platform	Medium	23
Level Data	RANDA	Medium	21
Velocity Network Foundation	Velocity Career Wallet	Medium	21
Instructure	Canvas Credentials	High	19
Arizona State University	Pocket Wallet	High	19
SpruceID	Credible and Verifier	High	19
Gobekli	Universal Talent Passports	High	18
Truvera (formerly Dock)	Truvera Dock Certs	High	18
Territorium	TerritoriumCLR	High	17
TrueCred	TrueCred	High	16
Digital Bazaar	Veres Wallet	High	16
Learning Economy Foundation	LearnCard	High	15
IQ4	IQ4 Wallet	High	13
Proof of Knowledge (POK)	POK	High	12

Source: LWYL Studio (forthcoming, 2026)

Centralized Control in Decentralized Clothing

Two technical innovations which promise increased individual holder (or decentralized control) Self-Sovereign Identity (SSI) and Decentralized Identifers (DIDs). They were developed to *prevent* exactly this kind of vendor lock-in. The original vision imagined a truly decentralized ecosystem where people could move freely between services. Instead, we've created what is called *'decentralized washing', systems that adopt the language and aesthetics of decentralization while recreating centralized power structures.*

Your digital wallet may feel like a personal vault, but it functions more like a hotel safe: *accessible only through the property's infrastructure, subject to the owner's rules, and ultimately controlled by someone else's master key.*

This is vendor lock-in, and it represents a fundamental betrayal of the self-sovereignty promise.

INSTITUTIONAL DEPENDENCE

It's not just individuals locked in, institutions can be too. Once a university, workforce board, or employer adopts a vendor's credentialing stack, they may:

▶ Have to continue paying for storage or API access

▶ Be unable to issue new credentials outside that system

▶ Rely on proprietary formats that are hard to translate or standardize

▶ Lose historical data if the vendor discontinues service

Switching costs are not just financial, they involve retraining staff with system knowledge, updating processes, and rebuilding integrations with HR systems, learning management systems (LMS), student information systems (SIS), and analytics tools.

THE FINANCIAL MODEL: WHO REALLY PAYS?

In vendor-driven ecosystems, there are several revenue models we have to consider. These include:

- **Per-credential fees**: Vendors charge institutions for each credential issued or verified.
- **Subscription models**: Platforms charge annual fees to maintain credentialing capacity.
- **Verification gateways**: Employers or third parties must pay to access or confirm a credential.
- **Analytics layers**: Credential data is analyzed and sold back to institutions or partners.

This creates incentives to issue credentials in high volume, monetize verification, and design features around institutional needs, not individual learner and worker needs and empowerment. In fact, individuals in addition to not having a voice at the design and development stage, also do not financially benefit from this model, despite it all requiring their participation, data, and credential profile to function and work.

When Individuals Pay

Despite claims of worker or learner control, and sometimes ownership over their data, individuals often end up having to pay to use the systems. This may include paying to:

- Access credentials (e.g., paywalls for transcripts or export)
- Verify credentials for job applications
- Use advanced features in wallets (e.g., analytics, alerts)
- Transfer credentials between platforms

Some vendors embed microtransaction models into wallets, charging per transaction or for *premium* usage tiers. Others monetize learner and worker data through analytics or behavioral insights and artificial intelligence LLMs. Even when the platform is free, individuals *pay* with their information, feeding systems that profit from profiling and engagement tracking.

'If you're not paying for the product, you are the product.', Popular Internet adage

SYSTEM FRAGILITY AND SUSTAINABILITY RISKS

Vendor and institutional lock-in also creates fragility. If a company is acquired, pivots its business model, or shuts down, we have to ask what then happens to the credentials it issued or managed? Who now has access to any data associated with them? And how might it impact learners and workers?

Example cases include:

- ▶ Badgr's acquisition by Instructure changed roadmap priorities
- ▶ Canvas discontinuation of the free version of the Canvas Badges platform for new badge issuers
- ▶ Digital credentialing startups folding after funding dries up
- ▶ Blockchain-based platforms losing community support or token value

All these highlight the risk of digital credentials becoming *orphaned*, valid in theory yet unverifiable in practice. In a public infrastructure model, this would be unacceptable. A bridge that disappears when a contractor closes would be considered a failure of governance.

Yet this risk is tolerated in verifiable digital credentialing across learning and work contexts.

INTEROPERABILITY IS THE BROKEN PROMISE

Many vendors claim interoperability yet implement only partial standards needed to support it.

▶ They provide partial support for Open Badges, yet not W3C VCs

▶ Use of DIDs yet limited verification across ecosystems

▶ Support for only one wallet app yet not others

▶ Difficult to fine or use export, share, and manage credentials functionality.

True interoperability requires not just technical adherence to standards, yet active collaboration, open governance, and *cross-platform testing*. Without it, interoperability becomes marketing, while lock-in continues to deepen.

THE ROLE OF PUBLIC FUNDING

Ironically, many vendor-driven systems are funded with public or philanthropic dollars: Workforce board grants that pay for proprietary software, Depart of Education or Labor innovation funds that subsidize platform licenses, and Philanthropic funds that fund pilot and tools with no long-term exit or sustainability plan. This raises ethical questions:

▶ Should public money fund credentialing infrastructure that isn't publicly governed, portable, or universally accessible?

▶ And if it does, should private vendors provide a share of profits-made from the use of public funding to the program or project implementation?

'Digital public infrastructure should be publicly governed, not licensed from private gatekeepers.' Digital Public Goods Alliance (2022).

TOWARD EXIT-ORIENTED DESIGN

To reduce lock-in and platform dependence, verifiable LWR implementers should prioritize:

- ▸ **Data portability**: Clear, documented export functionality and formats
- ▸ **Credential longevity**: Validity outside issuing platforms
- ▸ **Storage flexibility**: Support for multiple account or wallet applications
- ▸ **Verifier openness**: No paywalls for checking credentials
- ▸ **Interoperability by default**: Across VDC, and DID protocols

These principles must be embedded at the funding and design stage, not bolted on later as compliance checkboxes partially met.

SHARED GOVERNANCE AND PUBLIC STEWARDSHIP

One path forward is to treat credentialing infrastructure as a public utility. That means:

- ▸ Open-source reference implementations
- ▸ Credential registries operated by non-profits or public agencies
- ▸ Governance boards with seats for workers, learners, and community organizations providing quality assurance and compliance services
- ▸ Contracts that mandate ownership, openness, data portability and ethical use
- ▸ Affordable pricing and business models

The alternative is an ecosystem of fragmented, rent-seeking vendors who treat records of learning and work as assets to be mined and sold, not public and human goods to be protected.

CONCLUSION

Vendor lock-in and platform dependence are not just technical inconveniences, they are structural features of how much of today's credentialing infrastructure has been built.

Promises of decentralization, storage or wallet choice, and interoperability often mask deep power asymmetries, where private vendors control access, formats, and even the economic terms of participation.

When the infrastructure for recognizing learning and work behaves like a walled garden rather than a public utility, the risks ripple out to individuals, institutions, and entire ecosystems.

This chapter has shown that lock-in undermines the core values that verifiable digital credentials claim to advance: portability, autonomy, equity, and ownership. Learners and workers end up paying for access to their own records, institutions lose flexibility, and public dollars underwrite private systems that may not survive or remain accountable. Without intentional exit strategies, portability standards, and public-interest governance, digital verifiable credentials risk becoming another closed market rather than a liberating technology. The entire industry appears to conflate marketing with reality.

To realize the original vision of issuing learning and work records, as a public service that *empowers people*, not just platforms, designers, funders, and policymakers must insist on data portability, open standards, shared governance, and stewardship models that truly center the needs of learners and workers.

Anything less will recreate old dependencies in a new digital form.

FURTHER READING

1. Benjamin, R. (2019). *Race After Technology: Abolitionist Tools for the New Jim Code*. Polity Press.

2. Boninger, F., Molnar, A., & Murray, C. (2017). *Asleep at the Switch: Schoolhouse Commercialism, Student Privacy, and the Failure of Policymaking*. National Education Policy Center. https://nepc.colorado.edu/publication/schoolhouse-commercialism-2017

3. Eubanks, V. (2018). *Automating Inequality: How High-Tech Tools Profile, Police, and Punish the Poor*. St. Martin's Press.

4. Mozilla Foundation. (2023). *Privacy Not Included* [Buyer's Guide]. https://foundation.mozilla.org/privacynotincluded/

5. Digital Public Goods Alliance. (2022). *Digital Public Goods Charter*. https://www.dpgcharter.org/

6. 1EdTech Consortium. *Understanding Digital Credentials*. https://www.imsglobal.org/understanding-digital-credentials

7. Credential Engine. *Credential Transparency Description Language (CTDL)*. https://credentialengine.org/

8. World Wide Web Consortium (W3C). (2022). *Verifiable Credentials Data Model 1.1*. https://www.w3.org/TR/vc-data-model/

08

WHEN CREDENTIALING INCENTIVES UNDERMINE PUBLIC ACCESS

▶ Credentialing infrastructure prioritizes institutional needs over individual access

▶ Nonprofits face conflicts when acting as both advocates and vendors

▶ Funding rewards visibility over long-term value

▶ Market incentives undermine portability and interoperability

▶ Solutions require transparency, ethics, and community participation

Verifiable digital credentials are widely promoted as transformative equity tools, promising to empower individuals to demonstrate their skills more effectively, access better employment opportunities, and exercise meaningful control over their own learning and work records. Advocates argue that these systems can level the playing field, making qualifications more portable, verifiable, and accessible across contexts.

The vision is compelling: a future where credentials follow people throughout their lives, crossing institutional boundaries and opening doors that traditional credentials cannot.

Yet behind this promise, the systems that issue, store, and verify these digital credentials are often designed, funded, and governed in ways that prioritize institutional efficiency, market expansion, and investor returns rather than the actual needs and interests of credential holders themselves. Infrastructure decisions favor vendors and platforms. Funding flows toward initiatives that demonstrate scale and visibility rather than sustained impact. Governance structures center organizations rather than individuals or communities.

This chapter examines how the economic incentives embedded in digital credentialing ecosystems, particularly vendor-driven models, shape infrastructure development, policy decisions, and product design in ways that can inadvertently undermine public access, interoperability, and equity. It explores the tensions that arise when nonprofits serve dual roles as public advocates and commercial vendors, when market incentives discourage portability, and when credentialing tools designed to serve people become optimized instead for institutions and markets.

Even when equity is the stated goal, the structural dynamics of how these systems are built and sustained can produce outcomes that work against it.

THE ECONOMICS OF CREDENTIALING PLATFORMS

Verifiable digital credentialing systems emerge from complex economic relationships. They are shaped and sustained by multiple actors, each with distinct priorities:

▶ **Commercial software vendors** pursuing revenue growth and market share

▶ **Philanthropic funders and foundations** advancing specific policy agendas and theories of change

▶ **Public-sector agencies** operating under performance mandates and accountability frameworks

▶ **Institutions and organizations**, schools, employers, workforce boards, seeking operational efficiency and regulatory compliance

These actors, rather than individual learners or workers, often function as the primary customer base for credentialing platforms. Product development responds accordingly, with resources flowing toward features that serve institutional buyers:

- **High-volume credential issuance** that demonstrates scale and reach

- **Enterprise integrations** with human resources systems, student information systems, and learning management platforms

- **Administrative dashboards** for tracking, reporting, and compliance monitoring

- **Data analytics and insights** that create additional revenue streams or support institutional decision-making

By contrast, features that would directly benefit individuals, intuitive interfaces, clear explanations of credential value, true data portability, support for diverse or non-standard use cases, are frequently underfunded, deprioritized, or treated as secondary concerns.

The result is infrastructure optimized for institutional workflows rather than individual agency, even when those individuals are the stated beneficiaries of the system.

WHO PAYS? WHO PROFITS?

A substantial market has emerged around verifiable digital credentialing infrastructure, with multiple revenue streams flowing through the ecosystem:

- **Institutions** purchase vendor platforms to issue badges, certificates, and digital transcripts

- **Employers** pay for verification services to authenticate candidate credentials

- **Government agencies** fund demonstration projects, pilot programs, and system implementations through grants and procurement

- **Investors** seek returns from wallet applications, blockchain infrastructure providers, and credentialing platforms

Yet the individuals at the center of this ecosystem, learners and workers whose data, achievements, and labor generate the credentials being traded, are rarely positioned as customers or beneficiaries. Instead, they are expected to:

- **Download and maintain applications** with opaque privacy practices and terms of service

- **Manage credentials across incompatible systems** without clear pathways for consolidation or transfer

- **Share personal data for verification purposes** without meaningful transparency about data use, storage duration, or revocation rights

- **Navigate platforms and interfaces** designed primarily for institutional workflows rather than individual needs

In some cases, individuals also bear direct costs such as paying fees to access their own records, print official copies, or verify credentials to third parties. This is particularly common with educational transcripts and professional licenses issued by government agencies. The result is a paradox: *those who generate the value in credentialing systems often have the least control, receive the fewest benefits, and sometimes must pay for the privilege of participating.*

PRODUCT DEVELOPMENT FOLLOWS THE MONEY

Vendors respond to customer demand, and the paying customers are typically institutions and employers. This leads to a misalignment of priorities. Some examples include:

- Complex interfaces built for admin functionality, not individual usability
- Lack of multilingual or mobile-first design
- No support for narrative credentials or contextual storytelling
- Delayed or absent development of revocation or consent controls

In summary, features that institutions and employers ask for get built. Features that learners and workers as holders need often don't.

This isn't always malicious, it reflects how product roadmaps are shaped by business models and revenue streams. Yet it results in systems where convenience for issuers outweighs ease, usefulness, and affordability for credential holders.

METRICS THAT DRIVE MISALIGNMENT

The measures used to evaluate credentialing platforms reveal whose interests the systems are designed to serve. Vendors marketing their products and funders reporting on their investments typically highlight metrics that demonstrate institutional adoption and operational efficiency, such as:

- **Number of credentials issued** (volume as proxy for impact)
- **Number of institutions onboarded** (market penetration and network effects)
- **Time to verification** (technical performance and system speed)
- **Employer partnerships established** (ecosystem participation and buy-in)

- **Integration completeness** (compatibility with existing enterprise systems)
- **Cost per credential issued** (operational efficiency)

These metrics are attractive because they are easy to measure, demonstrate scale quickly, and align with institutional procurement decisions. They tell a story of growth, reach, and technical capability that appeals to boards, investors, and policymakers.

Conspicuously absent from most evaluation frameworks are metrics that would center the experiences and outcomes of credential holders themselves, such as:

- **Holder comprehension** (do people understand what their credentials mean and how to use them?)
- **Real-world usage rates** (are credentials actually being presented to employers, schools, or other verifiers?)
- **Support request volume and resolution time** (how often do people need help, and do they get it?)
- **Digital literacy impact** (do systems build capability or create barriers?)
- **Credential abandonment or non-use** (how many issued credentials are never accessed or shared?)
- **Holder-reported confusion, frustration, or privacy concerns** (what is the lived experience of navigating these systems?)
- **Equity of access across populations** (who succeeds with these tools and who is left behind?)

This measurement gap is not accidental, it reflects the underlying economic structure of credentialing ecosystems. Platforms are incentivized to optimize for what customers and funders will pay for and what procurement officers can justify, not necessarily for what individuals need or value. As Credential Engine's research revealed, while over one million unique credentials are now issued annually in the United States, there is very little systematic or reliable data on whether recipients understand them, use them, or find them valuable in advancing their education and career goals.

The industry has succeeded in building infrastructure for scale, but has largely failed to measure or prioritize whether that infrastructure actually serves the people it claims to empower.

THE ROLE OF PILOT FUNDING AND SHORT-TERM GRANTS

The lifecycle of credentialing infrastructure is fundamentally shaped by the funding mechanisms that bring it into existence. Most digital verifiable credentialing initiatives begin not as long-term public infrastructure investments, but as time-limited demonstration projects funded through competitive grants. Philanthropic foundations and public agencies typically structure these opportunities with:

- ▶ **Compressed timelines** of 6 to 18 months from award to final reporting

- ▶ **Predetermined outcome targets** emphasizing credential issuance volume, institutional partnerships, or platform adoption rates

- ▶ **Deliverables focused on visibility** such as conference presentations, white papers, case studies, and media coverage

- ▶ **Narrow eligibility criteria** that favor established organizations with existing technical capacity

- ▶ **Matching requirements** that disadvantage under-resourced communities and smaller institutions

These structural features create a predictable set of incentives that systematically distort project design and implementation. These include:

1. **A rush to deployment:** With tight deadlines and upfront deliverable commitments, teams prioritize speed over participatory design, skipping essential steps like human-centered design research, accessibility testing, or iterative prototyping with actual credential holders.

2. **Optimization for measurable outputs:** When success is defined by the number of credentials issued or institutions onboarded within the grant period, resources flow toward activities that inflate those numbers such as mass issuance campaigns, promotional events, bulk integrations, rather than toward the slower work of building understanding, trust, and sustainable adoption.

3. **Risk aversion in population selection:** Projects gravitate toward populations that are easier to reach, more digitally literate, and already engaged with existing systems. Harder-to-reach communities, disconnected youth, workers without reliable internet access, immigrants navigating language barriers, formerly incarcerated individuals rebuilding their lives, are often excluded not because they wouldn't benefit, but because serving them well would complicate timeline and metric achievement.

4. **Deferred attention to sustainability:** Questions about long-term governance, ongoing maintenance costs, data stewardship responsibilities, and post-grant technical support are frequently treated as 'future considerations' that can be addressed after the pilot proves successful. In practice, most pilots end when funding ends, leaving behind credentials that can't be verified,

platforms that are no longer maintained, and communities that invested time and trust with nothing to show for it.

This all results in misalignment in action. The consequences of these dynamics are not hypothetical, they are visible across the credentialing landscape:

- **Workforce credentialing programs** that issue thousands of digital badges to job seekers who never present them to employers, either because employers don't recognize or accept them, or because the wallet interfaces are too confusing to navigate without ongoing support that disappeared when the grant ended.

- **Educational badge ecosystems** where learners accumulate credentials without understanding what competencies they represent, how they differ from traditional transcripts, or when and how sharing them might advance their goals, resulting in digital accounts full of badges that are never opened after the initial issuance email.

- **Verification infrastructure** where product updates pushed by vendors break backward compatibility, requiring people to reauthenticate repeatedly or migrate credentials to new platforms, effectively punishing early adopters and discouraging continued engagement.

- **Blockchain-based credentialing pilots** that generate compelling demonstration videos and conference presentations during the funded period, but where the ongoing costs of maintaining nodes, paying transaction fees, or providing holder support prove unsustainable once the grant money is spent, leaving credentials inaccessible or unverifiable.

- **Interoperability initiatives** that successfully connect multiple platforms during the pilot phase, but fragment again when different vendors pursue incompatible product roadmaps after the collaborative funding incentive disappears.

These patterns are not exceptions or implementation failures, they are the predictable outcomes of a funding model that treats credentialing infrastructure as a series of discrete innovations to be demonstrated rather than as essential public infrastructure to be built, maintained, and governed for the long term.

The result is a landscape littered with abandoned pilots, disappointed learners and workers, and lessons learned that are rarely incorporated into the next round of grant-funded experiments or shared with others to learn from.

WHY PUBLIC INFRASTRUCTURE PRINCIPLES MATTER

Verifiable digital credentialing systems are frequently positioned as public goods or essential public infrastructure for lifelong learning, workforce mobility, and economic opportunity. They are promoted as mechanisms for equity, access, and empowerment. Government agencies fund them. Educational institutions adopt them. Workforce development programs depend on them.

Yet despite serving fundamentally public functions, these systems are overwhelmingly built, owned, and operated according to private-sector logics.

Infrastructure or Product?

The distinction matters. Public infrastructure, such as roads, bridges, water systems, public libraries, postal services, are designed with fundamentally different priorities than commercial products. The design priorities may include:

- ▶ **Universal access** as a design requirement, not a market segmentation strategy

- ▶ **Long-term sustainability** is planned from the outset, not deferred as a post-revenue concern

- ▶ **Public accountability** shapes governance, not shareholder value or investor timelines

- ▶ **Interoperability** is built in by design, not treated as a competitive disadvantage

- ▶ **Maintenance and stewardship** are ongoing commitments, not negotiable expenses

- ▶ **Governance structures** include mechanisms for public input, not just vendor roadmaps

Most credentialing platforms fail nearly every one of these tests:

Access is conditional and revocable: Institutions license credentialing platforms through annual contracts that can be discontinued, renegotiated, or transferred to different vendors. When contracts end or vendors change, credential holders may lose access to their own records, or discover that verification links no longer function. Unlike a birth certificate or Social Security card, imperfect as those systems may be, verifiable digital credentials often exist only as long as someone continues to pay the subscription fee.

Sustainability is not guaranteed: Platforms are built on business models that require continuous revenue, from licensing fees, transaction charges, verification services, or data monetization. If a vendor is acquired, pivots to a different market, or simply shuts down unprofitable product lines, the credentials disappear. There is rarely a legal requirement for data portability, sunset migration support, or long-term archival responsibility.

Governance is opaque: Decisions about platform features, data practices, privacy policies, pricing structures, and which credentials are supported are made by vendors and their enterprise customers, not by credential holders or the communities they serve. Terms of service change unilaterally. Privacy practices evolve without consent. There is no public comment period, no appeals process, no mechanism for democratic input.

Interoperability is a competitive threat: In a market-driven ecosystem, vendor lock-in is a feature, not a bug. Platforms that make it easy for institutions to switch providers or for individuals to export their credentials to competing wallets are disadvantaging themselves commercially. Standards exist, W3C Verifiable Digital Credentials, IMS Open Badges, various blockchain protocols, yet implementation is selective, and true portability across ecosystems remains elusive in practice.

Stewardship is absent: Who is responsible for ensuring that credentials issued today will still be accessible, verifiable, and meaningful in five years? Ten years? Twenty? In the current model, the answer is often no one. Pilots end. Funding expires. Vendors exit the market. Unlike public records systems with legal mandates for retention and access, digital credentials exist in a regulatory gray area with few enforceable obligations.

What Digital Public Infrastructure Would Require: If credentialing systems were genuinely designed as public infrastructure rather than commercial products, they would embody fundamentally different characteristics (Table 12).

Table 12: Essential Elements of Digital Public Infrastructure for Credentialing

Infrastructure Element	What It Requires	Why It Matters
Mandatory Interoperability Standards	Regulation or procurement requirements ensuring credentials can move across platforms without permission or penalty	Prevents vendor lock-in; enables true portability; protects credential holders from losing access when institutions switch providers
Open-Source Reference Implementations	Non-proprietary baseline platforms that any organization can adopt, adapt, or build upon	Prevents single-vendor control of the ecosystem; enables innovation; reduces costs; ensures transparency in how systems work
Public Governance Structures	Meaningful participation from credential holders, civil society organizations, and affected communities, not just institutions and vendors	Ensures systems serve peoples needs; provides accountability; creates mechanisms for democratic input on policies and priorities

Long-Term Sustainability Models	Public investment funding models not dependent on private capital seeking returns	Guarantees ongoing access and maintenance; removes pressure to monetize people's data; enables focus on public value over profit
Legal Frameworks for Data Rights	Enforceable rights for credential holders to access, export, correct, and delete their own data	Establishes individuals as primary actors; provides legal recourse when rights are violated; shifts power from platforms to people
Transparent Accountability Mechanisms	Public audits, impact assessments, and redress processes when systems fail or cause harm	Enables oversight; creates consequences for failures; builds trust; provides pathways to remedy when individuals are harmed
Equity Mandates	Requirements for accessibility, multilingual support, offline alternatives, and proactive outreach to marginalized populations as conditions of public funding	Ensures systems work for everyone, not just digitally privileged populations; prevents credentialing infrastructure from deepening existing inequities

The Digital Public Goods Alliance, in formalizing its charter in 2022, outlined principles for digital public infrastructure that credentialing systems rarely meet: *open licensing, platform independence, privacy protection, adherence to standards, and governance structures that prevent single points of control.* These principles remain aspirational in most verifiable and digital credentialing contexts.

The question is not whether digital credentials can be useful, they can. The question is whether they will be built as public infrastructure accountable to the people they claim to serve, or as private products optimized for institutional buyers and investor returns. Current trajectories suggest the latter, with predictable consequences for equity, access, and long-term sustainability.

THE VENTURE-BACKED CREDENTIALING ECONOMY

Record, credential, and wallet tools are increasingly backed by venture capital. These startups must grow fast, demonstrate market traction, and generate monetizable data flows. Their products are often offered 'free' to learners and workers yet they generate value by capturing usage data, monetizing analytics and locking people and institutions into proprietary formats. This extractive model is antithetical to the ideals of public credentialing, where *people should own their records, not trade them for access.*

EVEN NONPROFITS MAY NOT BE IMMUNE

It is tempting to assume that nonprofit organizations involved in digital credentialing ecosystems act as impartial stewards, operating above the market pressures and competing interests that shape commercial vendors. However, many have become increasingly embedded in the same incentive structures, accepting vendor-focused funding, providing paid consulting services, developing proprietary platforms, or operating in close partnership with for-profit entities. These entanglements raise critical questions about independence, transparency, and whose interests are ultimately being served.

The nonprofit and consortium-based organizations shaping verifiable digital credentialing infrastructure occupy influential positions, setting technical standards, operating credential registries, conducting research, convening stakeholders, and building implementation tools. Their work is often essential, filling gaps that neither government nor the private sector adequately address. Yet the line between public-interest stewardship and market participation has become increasingly blurred.

Understanding Dual Roles and Competing Incentives

The table below provides an overview of several prominent organizations operating in the U.S. and serving international credentialing ecosystem. Rather than offering evaluations or endorsements, it highlights each organization's primary contributions alongside reflective questions about governance structures, transparency practices, potential conflicts of interest, and alignment with public-interest values.

These questions are invitations to examine how power, funding, and decision-making operate within credentialing systems, and how these dynamics may affect equity, learner agency, and trust as verifiable digital credentials continue to scale. Organizations operating in complex, rapidly evolving ecosystems will inevitably face tensions between sustainability and mission, between responsiveness to funders and accountability to learners and workers, between technical innovation and equity considerations. *The challenge is whether those tensions are acknowledged, governed transparently, and resolved in ways that prioritize the interests of credential holders.*

Table 13: Nonprofit and Consortium Organizations in Digital Credentialing

Entity	Contributions	Reflective Questions
Education Design Lab (EDL)	Supports education pathway initiatives; develops frameworks for skills-based learning; operationally connected to microcredential infrastructure	▶ How are decisions about which pathways and credentials to promote made, and who participates in those decisions? ▶ What is the relationship between advocacy for microcredentials and partnerships with microcredential platform providers? ▶ How does operational involvement in credential infrastructure affect the organization's capacity to evaluate those systems critically?
Competency-Based Education Network (C-BEN)	Convenes institutions; provides guidance on competency-based education implementation; supports system development	▶ How are competency frameworks evaluated for equity implications, particularly for learners from non-traditional backgrounds? ▶ When providing implementation support to institutions, how is success defined, by institutional adoption or by learner outcomes? ▶ How does governance ensure that competency-based approaches serve learners rather than simply reducing costs or accelerating completion?

Credential Engine (CE)	Operates the Credential Registry and Credential Transparency Description Language (CTDL); increases visibility across credential offerings	▸ How does neutrality function when credentials of varying rigor, recognition, and labor market value are presented side by side without quality differentiation? ▸ How can metadata standards and partnerships remain centered on learner understanding and real-world value rather than scale and vendor adoption alone? ▸ Who decides what counts as a credential worthy of inclusion, and what governance mechanisms exist for those decisions?
Digital Promise Global (DPG)	Conducts edtech research; advocates for equity; offers credentialing platforms and consulting services to schools and organizations	▸ How are evaluative research, equity advocacy, and fee-based service delivery roles clearly delineated? ▸ What safeguards ensure that research and validation efforts remain independent when the organization also sells credentialing services? ▸ How are potential conflicts of interest disclosed and managed when conducting research on markets where the organization is also a vendor?

Digital Credentials Consortium (DCC)	Develops open-source tools (Learner Credential Wallet); promotes learner-controlled verifiable credentials; advances university adoption	▶ How does a governance model anchored in higher education institutions account for non-degree learners, community college pathways, and workforce credentials? ▶ How are learner voice, accessibility needs, and equity considerations incorporated as tools scale beyond founding member institutions? ▶ What mechanisms ensure that open-source tools remain genuinely accessible to under-resourced organizations, not just well-funded universities?
IMS Global / 1EdTech	Maintains Open Badges, Comprehensive Learner Record (CLR), Learning Tools Interoperability (LTI), and other widely adopted edtech standards	▶ How does a membership-driven model ensure inclusion of perspectives beyond well-resourced institutions and established vendors? ▶ How are social, ethical, and equity implications, such as consent, power dynamics, and data rights, addressed alongside technical interoperability? ▶ What governance structures exist to prevent standards from being shaped primarily by the largest paying members?

Jobs for the Future (JFF)	Produces influential research reports; hosts national convenings; facilitates pilots such as Plugfest and Digital Wallet market scans	▶ When research, convening, and promotion of specific credentialing tools occur within the same organization, how are evaluation criteria defined and applied consistently?
		▶ How is neutrality maintained when some vendors or approaches receive more visibility through JFF platforms than others?
		▶ How are community voices, particularly those of credential holders, incorporated into research design and review processes?
Learning Economy Foundation (LEF)	Supports credential initiatives and workforce pathways; operationally connected to LearnCard and related open-source tooling developed and owned by for-profit entity WeLibrary LLC	▶ How are governance decisions made when nonprofit leadership overlaps with for-profit technical development entities?
		▶ How are public-interest goals balanced with market incentives when tools are owned by a separate for-profit company?
		▶ What transparency mechanisms exist regarding financial relationships, licensing arrangements, and decision-making authority between the nonprofit and for-profit entities?

U.S. Chamber of Commerce Foundation (USCCF) T3 Network	Develops frame-works for skills trans-parency; convenes employers; builds implementation tools and prod-ucts for workforce credentialing	▸ When standards-setting, employer convening, and product development coexist within the same entity, how are governance and public-interest safeguards maintained?
		▸ How are worker agency, privacy, and equity considerations balanced against employer efficiency and hiring priorities?
		▸ What mechanisms ensure that skills frameworks reflect worker interests and mobility, not just employer workforce management needs?

Moving Toward Accountability

These organizations are contributing important work to education and workforce transformation. The challenge they collectively illustrate is the difficulty of maintaining impartiality and public accountability when funding structures, service models, technology development, and strategic alliances overlap in complex ways.

None of these dynamics are inherently disqualifying. Organizations need sustainable funding. Expertise in research does not preclude expertise in implementation. Developing tools can inform standards work, and vice versa. The critical question is whether these relationships are governed transparently, whether conflicts of interest are acknowledged and managed, and whether decision-making structures genuinely center the interests of learners and workers rather than institutions, funders, or vendors.

Transparency, independent governance, community participation, and commitment to equity are not obstacles to innovation, they are prerequisites for building credentialing systems that serve the public interest rather than simply the organizations that build and benefit from them.

IMPLICATIONS FOR TRUST AND PUBLIC GOVERNANCE

When nonprofit organizations and consortiums simultaneously serve as researchers, advocates, standard-setters, and service providers, or operate in close partnership with for-profit entities without clear structural separation, several risks to public trust and equitable outcomes emerge:

▶ **Compromised independence:** Research findings, policy recommendations, and 'best practices' guidance may be shaped, consciously or unconsciously, by financial relationships, partnership commitments, or organizational revenue models. When the organization evaluating credentialing systems is also selling credentialing services, or when standards bodies are funded primarily by the vendors implementing those standards, the credibility of claims about effectiveness, equity, or user benefit is fundamentally undermined.

▶ **Monetization of influence:** Advisory roles, policy convenings, and thought leadership platforms can translate directly into market advantage for organizations that also operate as vendors or maintain close ties to specific technology providers. Invitations to present at conferences, contribute to policy briefs, or participate in government working groups become marketing opportunities disguised as public service. The line between sector leadership and competitive positioning becomes impossible to discern.

- ▸ **Exclusion of affected communities:** When governance structures, research agendas, and success metrics are defined primarily by funders, institutional members, and vendor partners, the voices of credential holders themselves, learners, workers, job seekers, are typically absent or tokenized. Priorities reflect what can be funded and what serves institutional buyers, not what communities most affected by these systems actually need or want. Human-centered design becomes a consulting service purchased by institutions, not a genuine transfer of power to people such as holders and the organizations who serve them.

- ▸ **Erosion of accountability:** Without independent evaluation, transparent governance, and mechanisms for redress when systems cause harm, there is no meaningful accountability. Organizations can claim public-interest missions while operating according to market logics. Failures are reframed as 'lessons learned' in the next grant proposal rather than addressed through structural change or consequences for those responsible.

- ▸ **Legitimacy crises:** Digital credentialing systems, especially emerging verifiable systems built within these entangled ecosystems risk losing public trust, not because the underlying technology is necessarily flawed, but because the governance is opaque, the incentives are misaligned, and the beneficiaries are unclear. When people cannot distinguish between genuinely independent guidance and vendor-influenced promotion, skepticism becomes the rational response. When communities see their concerns ignored while institutional buyers drive product roadmaps, adoption becomes resistance.

Pathways to Accountability

Rebuilding and maintaining trust requires more than good intentions. It demands structural commitments that create real accountability. Some examples include:

▸ **Financial transparency:** Full disclosure of funding sources, vendor partnerships, licensing arrangements, consulting contracts, and any financial relationships that could create conflicts of interest. This information should be publicly accessible, regularly updated, and presented in plain language, not buried in footnotes or annual reports.

▸ **Structural separation of roles:** Clear boundaries between research and service delivery, between standards-setting and implementation, between evaluation and promotion. Public focused organizations should not simultaneously develop tools, sell those tools, conduct research on those tools, set standards for the tools and advocate for policies that would mandate those tools. When multiple roles are necessary, they should be governed by separate entities with independent boards and transparent decision-making processes.

▸ **Participatory governance:** Meaningful representation of credential holders, not just institutions that issue credentials, in governance structures, research design, priority-setting, and evaluation criteria. This means compensating community members for their expertise, creating decision-making power (not just advisory roles), and building accountability mechanisms that give affected communities real recourse when systems fail them.

▸ **Independent evaluation:** Third-party assessment of credentialing systems, conducted by researchers and evaluators with no financial stake in the outcomes. Evaluation criteria should be defined with input from credential holders and should measure impact on their lives, not just adoption rates, issuance volume, or institutional satisfaction

▶ **Public ownership of essential infrastructure:** Critical elements of credentialing systems, registries, verification protocols, identity standards, should be governed as public infrastructure with open-source implementations, not as proprietary assets controlled by nonprofits that depend on licensing fees for sustainability.

These commitments are not impossible. They require choosing mission over revenue growth, accountability over influence, and long-term public trust over short-term funding opportunities. Organizations that cannot make these choices may be contributing valuable work, but they are not functioning as impartial stewards of public infrastructure.

Recognizing that distinction is essential for anyone trying to build credentialing systems that genuinely serve the public interest.

REALIGNING INCENTIVES: FROM INSTITUTIONAL EFFICIENCY TO PUBLIC ACCOUNTABILITY

The misalignments documented throughout this chapter are not inevitable, they are the products of specific design choices, funding structures, and governance models that can be changed. Shifting credentialing ecosystems toward genuine public benefit requires deliberately restructuring the incentives that shape how platforms are built, evaluated, funded, and sustained.

The following table outlines concrete strategies for realigning incentives away from institutional efficiency and market growth, and toward equity, usability, and accountability to credential holders. These are not aspirational principles, they are actionable policy interventions that funders, government agencies, procurement officers, and governance bodies can implement immediately. Each strategy addresses a specific dysfunction in current practice and offers a clear alternative that centers the needs and rights of the people these systems claim to serve.

Realignment will not happen through voluntary adoption or market competition alone. It requires deliberate intervention: grant requirements that enforce different priorities, procurement standards that reward different behaviors, governance structures that shift power, and regulatory frameworks that establish credential holders' rights as non-negotiable. *The question is not whether better alternatives exist, but whether institutions with the power to demand them will choose to do so.*

Table 14: How Might We Realign Incentives

Strategy	Description
Fund usability, not just issuance	Grants should require evidence of usability testing with diverse populations, accessibility compliance, inclusive design audits, and documented holder-centered improvements, not just credential volume or institutional adoption.
Incentivize impact over volume	Success metrics should prioritize verified real-world usage, employment and education outcomes, holder satisfaction and comprehension, and equity of access, not simply the count of credentials issued or institutions onboarded.

Require portability and revocation rights	All publicly funded credentialing systems must enable credential holders to export, transfer, delete, or refuse credentials without penalty, with these rights established as preconditions of funding rather than optional features.
Include credential holders in governance	Boards, advisory groups, standards committees, and feedback mechanisms must include compensated representation from workers, learners, and frontline communities, with decision-making power, not just consultation roles.
Mandate ethical business models for vendors	Procurement processes should explicitly favor and reward vendors that demonstrate transparency in data practices, commitment to open standards and interoperability, non-exploitative pricing, and accountability to end users, not just lowest cost or fastest deployment.

CONCLUSION

Verifiable digital credentials hold genuine potential. At their best, they can help people make skills visible across contexts, navigate fragmented education and labor systems, and exercise meaningful control over their own records. The vision of portable, trustworthy credentials that follow people throughout their lives, across jobs, institutions, and geographies, remains compelling and worth pursuing.

Yet, potential is not destiny. The infrastructure being built today will either advance that vision or betray it, and the difference lies not in the elegance of the technology but in the integrity of the systems we build that govern it.

The Cost of Misaligned Incentives

This chapter has documented how economic incentives embedded in credentialing ecosystems systematically distort outcomes. When business models reward credential volume over credential value, platforms optimize for issuance rather than use. When success is measured by institutional adoption rather than individual outcomes, products serve buyers instead of holders. When funding cycles prioritize visibility over sustainability, systems are built to impress funders during pilot periods and abandoned when grants end. When nonprofits blur the lines between research, advocacy, and service delivery, independence erodes and conflicts of interest multiply.

These are not isolated failures or implementation mistakes, they are predictable consequences of building public infrastructure according to private-sector logics. The harm produced is not always visible in conference presentations or quarterly reports, but it is real: *credentials that confuse rather than clarify, systems that exclude rather than include, platforms that disappear when the funding runs out, and communities that invest trust only to be disappointed when institutions move on to the next pilot.*

What Accountability Requires

Public trust in digital credentialing infrastructure cannot be assumed or marketed, it must be earned through demonstrated commitment to the people these systems claim to serve. This requires more than adding 'equity' to mission statements or conducting holder research as a grant deliverable. It demands structural accountability demonstrated through:

- **Transparency** about who funds, governs, profits from, and controls credentialing systems, not just in annual reports, but in plain language accessible to credential holders themselves.

- **Governance structures** that give credential holders decision-making power, not just advisory roles or holder testing sessions, with compensation for their expertise and real mechanisms for redress when systems cause harm.

- **Business models** that do not depend on extracting value from the people they claim to empower, through surveillance, data monetization, or requiring individuals to pay for access to their own records.

- **Standards and regulations** that establish portability, interoperability, and data rights as non-negotiable requirements, not optional features vendors can choose to implement or ignore.

- **Funding mechanisms** that reward long-term sustainability, equity of access, and demonstrated impact on credential holders' lives, not just speed to market, institutional adoption rates, or impressive demonstration videos.

- **Independent evaluation** conducted by researchers with no financial stake in the outcomes, using metrics defined with input from affected communities, measuring what matters to people's lives rather than what is easy to count.

The Choice Ahead

The infrastructure decisions being made now, about standards, governance, funding, and business models, will shape credentialing systems for decades. Those decisions are being made largely by institutions, vendors, and funders, with credential holders themselves relegated to the roles of research subjects, pilot participants, and people using them whose feedback may or may not be incorporated.

This does not have to continue. Government agencies can change procurement requirements. Funders can re-structure grant criteria. Standards bodies can reform governance. Institutions can demand portability. Nonprofits can choose mission over revenue. Regulators can establish credential holders' rights. Communities can refuse to participate in systems that do not serve their interests.

If digital credentials are to fulfill their promise as tools for equity and opportunity rather than becoming yet another mechanism for surveillance, exclusion, and institutional control, then the systems that issue, store, and verify them must be rebuilt according to different principles. Not innovation for its own sake, but infrastructure *built with care*. Not disruption as branding, but accountability as practice. Not moving fast and breaking things, but building slowly, deliberately, and in genuine partnership with the communities these systems will affect.

The technical standards exist. The policy frameworks are being articulated. The evidence of what does and does not work is accumulating. What remains uncertain is whether those with the power to demand better systems, funders, policymakers, institutions, and communities, will choose to exercise that power.

The credentials we build will reflect the values we encode. The question is not whether the technology works, but who it works for, and who gets to decide.

FURTHER READING

1. Credential Engine. (2022). *Counting U.S. postsecondary and secondary credentials*. https://credentialengine.org/all-resources/counting-credentials/

2. Digital Public Goods Alliance. (2022). *Digital Public Goods Charter*. https://www.dpgcharter.org

3. Eubanks, V. (2018). *Automating inequality: How high-tech tools profile, police, and punish the poor*. St. Martin's Press.

4. 1EdTech Consortium & Credential Engine. (2022). *Badge count 2022: Findings*. https://content.1edtech.org/badge-count-2022/findings

5. JFF. (2022). *Building a skills-based talent marketplace: Digital wallets*. https://info.jff.org/digital-wallets

6. JFF. (2024). *Jobseekers want digital credentials for skill-sharing: Are employers ready?* https://jff.org/jobseekers-want-digital-credentials-for-skill-sharing-are-employers-ready

7. Mozilla Foundation. (2023). *Privacy not included buyer's guide*. https://foundation.mozilla.org/en/privacynotincluded/

8. Rework America Alliance. (2020). *Skills-based sourcing and hiring resources*. Markle Foundation. https://www.markle.org/alliance-resources/

9. W3C. (2022). *Verifiable credentials data model v1.1* (W3C Recommendation). https://www.w3.org/TR/vc-data-model/

09

USABILITY, DESIGN, AND DIGITAL EXCLUSION

CHAPTER INSIGHTS

▶ How digital systems exclude vulnerable populations through design and access barriers

▶ Why wallet and app interfaces fail people: common human-centered design mistakes

▶ Building for everyone: accessibility considerations across disability, literacy, and language

▶ When technology conflicts with culture: addressing mismatches in language and values

Building functional Learning and Work Records (LWRs) issued as Verifiable Digital Credentials (VDCs) requires more than sound technical architecture and aligned incentives, it demands that systems work for the people who must use them. Yet a troubling gap exists between the empowerment narrative promoted by credential advocates and the frustrating experiences of many credential holders. Confusing interfaces, inaccessible designs, and assumptions about digital literacy create barriers that effectively lock out significant portions of the intended population.

This chapter investigates who gets left behind in digital credentialing systems, especially emerging verifiable credential systems, and why, examining the design flaws, accessibility failures, and infrastructure gaps that transform tools meant to democratize opportunity into new mechanisms of exclusion.

The scale of this challenge is substantial. Nearly one-third of U.S. workers lack foundational digital skills, yet 92% of jobs now require some level of digital competency (National Skills Coalition & Federal Reserve Bank of Atlanta, 2023). Meanwhile, approximately 15% of the global population experiences some form of disability, many of whom face compounded barriers when digital systems fail to meet basic accessibility standards. These statistics reveal more than individual struggles with technology, they expose system-ic patterns of exclusion where credential systems inadver-tently replicate existing inequities. When credential wallets and platforms require smartphone ownership, high-speed internet, and fluency in technical jargon, they effective-ly create *digital redlining* that mirrors historical patterns of discrimination in access to education and employment opportunities.

Beyond accessibility and digital literacy, cultural and lin-guistic barriers further complicate adoption. Credentialing systems often embed assumptions about language profi-ciency, cultural context, and familiarity with Western edu-cational models that alienate international holders, immi-grants, and English language learners. Interface metaphors drawn from American professional culture, such as portfoli-os, transcripts, and endorsements, may carry little meaning for people from different educational traditions or profes-sional contexts.

Similarly, the visual design, navigation patterns, and in-formation architecture of credential platforms typically re-flect the preferences and mental models of their developers or institutional administrators, rather than the diverse pop-ulations of learners and workers they aim to serve. When credential systems fail to account for this diversity, they don't simply inconvenience holders, they actively exclude entire communities from participation in the credentialing ecosystem.

Understanding why usability matters begins with rec-ognizing that even the most technically sophisticated cre-dential system fails if people cannot, or will not, engage with it.

WHY USABILITY MATTERS IN DIGITAL CREDENTIALING

The concept of usability emerged from the field of human-computer interaction in the 1980s, when researchers like Jakob Nielsen and Donald Norman began systematically studying how people actually interact with technology. Norman's influential work, *The Design of Everyday Things (1988)*, introduced the principle that when people struggle with a product, the fault lies not with the person but with the design. This insight fundamentally shifted how technologists approach system development, from expecting people to adapt to machines, to designing machines that adapt to people. In the decades since, usability principles have become standard practice in consumer technology, where even minor friction in the experience can determine whether a product succeeds or fails in the marketplace.

Yet credentialing systems have largely failed to embrace these lessons. Too often, credential platforms are designed by technologists for technologists, with little consideration for the diverse populations who must actually engage with them. The result is systems that prioritize technical sophistication over human experience, assuming levels of digital literacy, device access, and cultural familiarity that large segments of the population simply do not possess.

Usability in credentialing is not a luxury or aesthetic preference, it is a fundamental requirement for system viability. If a learner or worker cannot easily complete basic tasks, or an administrator has to spend hours navigating multiple screens and functions to issue or manage credentials, the system has failed regardless of how secure, private, or standards-compliant it may be.

The entire purpose of verifiable digital credentials should be to simplify access to learning and work records and facilitate their use in connecting people with opportunities. Essential credential interactions in this include:

- **Understanding** what a credential represents or signals to people and why it has value

- **Accessing** credentials from available devices (smartphones, tablets, computers)

- **Storing** credentials in a way that feels secure and retrievable for the holder and administrator

- **Sharing** credentials with employers, educational institutions, or other verifiers

- **Managing** credentials for a holders needs and so they may learn from the data held within them, and share, use, and export them as they wish

- **Integrating** credentials into application systems, enrollment processes, or professional profiles

- **Updating** credentials when information changes or new achievements are earned

- **Troubleshooting** when something goes wrong, with clear help resources or access to responsive customer support

Yet poorly designed systems often achieve the opposite, creating new barriers that compound existing inequities. When credential holders must navigate confusing interfaces, decipher technical jargon, or troubleshoot compatibility issues, the system becomes another obstacle rather than an enabler, transforming what should be a tool of empowerment into yet another mechanism that privileges those who already possess technical expertise and resources.

Ultimately, usability determines whether digital credentials serve their intended purpose of democratizing access to opportunity, or simply create a new digital divide where only the technologically privileged can participate in the credentialing ecosystem.

THE REALITY OF DIGITAL LITERACY GAPS

The digital divide is not merely about access to technology, it extends to the capacity to use it effectively and how one needs to. Research from Pew Research Center reveals that a significant portion of Americans struggle with tasks that credential systems take for granted. In their 2016 study on digital readiness, Pew found that only 17% of adults qualify as 'digitally ready,' possessing high confidence in using digital tools, comfortableness with new technology, and the ability to determine trustworthiness of online information (Pew Research Center, 2016). The remaining 83% fall along a spectrum from cautious to unprepared, with many not having the skills needed to navigate complex digital systems. By 2019, Pew's research on digital knowledge showed that Americans could correctly answer only 5 out of 10 questions about basic technology concepts, with substantial gaps in understanding security features like two-factor authentication, a technology increasingly embedded in credential systems (Pew Research Center, 2019).

The National Skills Coalition's comprehensive analysis paints an even starker picture. Their 2021 report found that nearly one in three U.S. workers lack foundational digital skills, the basic abilities required to function in today's digital economy (National Skills Coalition, 2021). This includes 13% who have no digital skills at all, and another 18% with only some capabilities. These individuals struggle with tasks that credential platforms assume as baseline competencies such as using a web browser effectively, navigating file systems, understanding cloud storage as well as export/import, distinguishing legitimate websites from fraudulent ones, managing passwords, and troubleshooting when applications malfunction. Yet these same workers, already disadvantaged in the digital landscape, are now being asked to manage digital verifiable credentials, understand the concept of digital wallets, make decisions about data sharing and privacy settings, and navigate the technical complexities of decentralized identity systems.

The gap is not distributed evenly across the population. Older adults face particular challenges, with those over 65 scoring significantly lower on digital literacy assessments and expressing less confidence in their ability to learn new technologies. Low-income workers often experience inconsistent access to devices and reliable internet, forcing them to attempt complex digital tasks on smartphones with small screens and limited data plans, or in public libraries with time constraints and competing demands on shared computers. Rural residents contend with bandwidth limitations that make loading credential platforms frustratingly slow or impossible, while also having fewer opportunities to develop digital skills through workplace technology exposure. English language learners confront interfaces designed exclusively in English, with technical terminology that challenges even native speakers, and no accommodation for different levels of language proficiency or cultural familiarity with digital credential conventions.

The consequences of these gaps extend beyond individual inconvenience. When credential systems require digital literacy that a third of the workforce doesn't have, they effectively create a credentialing separateness, a two-tiered system where digitally fluent individuals can document and leverage their learning achievements while those without such skills remain invisible in the credentialing ecosystem, unable to demonstrate their competencies regardless of their actual knowledge or capabilities. This exclusion perpetuates and amplifies existing inequities, as the populations most likely to struggle with digital skills are also those who would benefit most from diverse and digital credentialing pathways: workers without degrees, individuals seeking to transition careers, immigrants with international credentials, and people in underserved communities where educational and economic opportunities, as well as a digital access, are already limited.

WALLET AND PLATFORM INTERFACES: BUILT FOR WHOM?

Credential wallets and platforms serve as the primary gateway through which individuals access, manage, and share their learning and work records within digital credentialing ecosystems. Yet a fundamental disconnect exists between the populations these systems claim to serve and the actual design choices that shape wallet and platform interfaces.

An examination of current credential wallet implementations reveals a troubling pattern: *most are engineered by technical specialists, tested by early adopters with significant digital expertise, and optimized for the preferences and capabilities of developers and administrators rather than the general population who must ultimately use them.*

This design bias manifests in multiple ways that create systematic barriers to adoption:

► **Insufficient Onboarding and Support** Most credential wallets and platforms provide minimal guidance during the crucial initial setup phase. People are expected to understand complex concepts with little explanation of what these terms mean or why they matter. Tutorial content, when it exists, often assumes baseline technical knowledge that many holders don't have. Help documentation is frequently written in technical language that mirrors developer documentation rather than addressing the practical questions non-technical holders have such as: 'Where are my credentials actually stored?' 'What happens if I lose my phone?' 'Who can see or access this information?'

▶ **Impenetrable Technical Language** Wallet and platform interfaces routinely deploy terminology drawn from cryptography, software development, and identity protocols without translation into plain language. Holders encounter phrases like 'cryptographic proof,' 'verifiable presentation,' 'JSON-LD format,' 'data integrity suite,' 'selective disclosure,' and 'zero-knowledge proof' with little context for understanding what these terms mean or how they affect credential functionality. Even seemingly simple actions are obscured by jargon: 'issuing a verifiable credential,' 'verifying with a trust registry,' or 'rotating cryptographic keys.' For populations already uncertain about digital systems, this language creates cognitive barriers that signal the system is not designed for them.

▶ **Developer-Centric Interface Design** The visual and interaction design of credential wallets and platforms also often reflects the preferences of their creators rather than evidence-based research on how diverse populations interact with digital systems. Interfaces prioritize information density and technical functionality over clarity and ease of use. Navigation structures assume familiarity with developer tools and technical applications. Information architecture reflects how credentials work technically rather than how holders think about their achievements and qualifications during life events. The result is systems that feel foreign and intimidating to anyone outside the technical community that designed them.

- **Inadequate Accessibility Considerations** Many credential wallets fail to meet even basic accessibility standards, effectively excluding people with disabilities from participation. Common failures include insufficient color contrast that makes text difficult to read for people with low vision, lack of screen reader support that prevents blind holders from navigating the interface, absence of keyboard navigation for those who cannot use pointing devices, small touch targets that create challenges for people with motor disabilities, and time-limited interactions that disadvantage people who process information more slowly. These are not obscure edge cases—approximately 15% of the global population experiences some form of disability—yet accessibility is routinely treated as an afterthought rather than a core design requirement.

- **Inadequate Testing with Representative Populations** Perhaps most fundamentally, credential wallets are typically tested exclusively with early adopters, technical enthusiasts, and developers—populations with high digital literacy, strong motivation to engage with emerging technology, and technical support networks to help troubleshoot problems. These testing populations cannot reveal the friction points that will frustrate and ultimately exclude the broader population. When designers never observe a low-literacy holder struggling to understand what a "verifiable credential" is, or a person with limited English proficiency attempting to decipher technical instructions, or an older adult trying to complete setup on an unfamiliar smartphone, those pain points remain invisible and unaddressed.

The consequences of these design failures are predictable and measurable. Credential holders abandon the wallet setup process partway through when confronted with confusing instructions or overwhelming technical complexity. Those who successfully complete setup often fail to understand how to retrieve or share their credentials when opportunities arise. Others complete the technical steps but lack confidence in the system's security or trustworthiness, leading them to avoid using it for important transactions. Still others successfully receive credentials but cannot figure out how to integrate them into job applications or educational enrollment processes, rendering the credentials functionally useless despite their technical validity.

These outcomes represent not merely poor design but a fundamental failure to center the needs and capabilities of the populations that credentialing systems claim to serve. When wallet interfaces privilege technical sophistication over usability, they perpetuate a system where credentials function primarily as signals within an elite technical community rather than practical tools for economic mobility and educational access across the full spectrum of society.

ACCESSIBILITY: BEYOND USABILITY TO UNIVERSAL DESIGN

While usability focuses on making systems easy and efficient for the general population, accessibility addresses whether people with disabilities, diverse abilities, and varying circumstances can use systems at all. This distinction matters: a credential wallet or platform might be highly usable for a sighted person with high-speed internet but completely inaccessible to someone who is blind, uses assistive technology, or has unreliable connectivity. Accessibility is not simply an extension of good usability, it is a prerequisite for inclusive credentialing that many current systems fundamentally fail to achieve.

The relationship between usability and accessibility is hierarchical: accessibility establishes the baseline for participation, while usability determines the quality of that participation.

A credential wallet with poor usability frustrates holders and reduces efficiency; a wallet with poor accessibility excludes entire populations from participation entirely. Yet credentialing systems routinely conflate these concepts, treating accessibility as an advanced feature rather than a foundational requirement, or assuming that 'easy to use' automatically means 'accessible to all.' Here are some things to consider to ensure accessibility for a diversity of credential holders.

Language and Linguistic Accessibility

The vast majority of credential wallets and platforms are available exclusively in English, despite the United States having over 67 million people who speak a language other than English at home. This monolingual design immediately excludes many immigrants, refugees, and English language learners from accessing their own credentials. Even when translation is offered, it is often implemented through automated services that produce awkward or incomprehensible technical language, or limited to a handful of high-resource languages while ignoring the dozens of other languages spoken in diverse communities. Linguistic accessibility extends beyond mere translation, it requires culturally appropriate metaphors, writing at appropriate literacy levels, and recognition that technical concepts may not translate directly across linguistic and cultural contexts.

Standards Compliance and Assistive Technology

The Web Content Accessibility Guidelines (WCAG 2.1) provide internationally recognized standards for making digital content accessible to people with disabilities, yet compliance among credential platforms remains remarkably low. According to research by the Center for Democracy & Technology (2021), fewer than 30% of credentialing and workforce development tools demonstrated robust accessibility features, despite many receiving public funding that technically requires accessibility compliance. Common violations include images without alternative text descriptions for screen readers, form fields without proper labels that prevent assistive technology from identifying input requirements, dynamic content updates that screen readers cannot detect, keyboard navigation failures that trap people who cannot use a mouse, color-coded information without text alternatives for colorblind individuals, and video content without captions or transcripts for deaf or hard-of-hearing holders.

These failures have concrete consequences. A blind credential holder using a screen reader may be unable to navigate the wallet interface because buttons lack descriptive labels. Someone with motor disabilities who relies on keyboard navigation may find themselves unable to access dropdown menus or modal dialogs designed only for mouse interaction. A deaf individual may miss critical audio alerts or verbal instructions in tutorial videos, while someone with dyslexia may struggle with dense text blocks and lack of visual hierarchy that would aid comprehension.

Device and Platform Assumptions

Credentialing platforms routinely make assumptions about the devices and platforms holders will use, creating accessibility barriers for those whose technology access differs from designer expectations. Some systems assume desktop or laptop computers with large screens, full keyboards, and high processing power or storage, failing to account for the reality that low-income populations disproportionately rely on smartphones as their primary or only internet-enabled device. Conversely, mobile-first designs may create small touch targets that are difficult for people with motor disabilities to activate accurately, or compress information in ways that challenge people with cognitive disabilities who need more visual space and clearer organization.

Platform assumptions extend to operating systems and browsers as well. Credential wallets that function only on recent iOS or Android versions exclude people using older devices they cannot afford to replace. Systems that require specific browsers disadvantage people using assistive technology that integrates better with particular platforms. Applications that demand high-bandwidth connections or significant data downloads create barriers for rural residents with limited connectivity or people managing restrictive data plans.

Connectivity and Offline Accessibility

The assumption of constant, high-speed internet access reflects a privileged perspective that ignores the reality of digital infrastructure inequality. According to the Federal Communications Commission, approximately 19 million Americans lack access to broadband internet, with rural areas and tribal lands particularly underserved. Yet most digital credential systems require persistent internet connectivity to access, verify, or share credentials. This design choice effectively excludes people with intermittent connectivity from participating in time-sensitive opportunities like job applications or enrollment deadlines.

Offline functionality, the ability to access and share credentials without an active internet connection, represents a critical accessibility feature that few digital credential systems implement. This capability matters not only for people in areas with poor connectivity, but also for situations where internet access is temporarily unavailable such as during power outages, in basement offices or rural areas with poor cellular reception, when traveling internationally with limited data, or when data plans have been exhausted before the end of the billing cycle. The failure to provide offline modes particularly disadvantages low-income holders who may have inconsistent access to connectivity and cannot afford to miss opportunities because their credentials are inaccessible at crucial moments.

Physical and Print Alternatives

The push toward exclusively digital credentials overlooks the reality that some populations prefer, need, or are legally entitled to physical documentation. Older adults who are uncomfortable with digital systems, people with disabilities that make digital interaction challenging, and those without consistent device access may require paper-based alternatives. Yet many digital credential systems increasingly offer no mechanism for generating physical versions of credentials, or produce versions that lack the same verification capabilities as their digital counterparts, creating a two-tiered system where those who cannot or choose not to use digital credentials receive inferior documentation.

Cognitive and Neurodiversity Considerations

Accessibility for people with cognitive disabilities, learning differences, and neurodivergent conditions requires distinct design considerations that credential systems rarely address. People with ADHD may struggle with multi-step processes that lack clear progress indicators or allow distraction. Those with autism spectrum disorders may find ambiguous language or social cues confusing, requiring more explicit and literal instructions. People with dyslexia benefit from specific font choices, adequate spacing, and alternatives to text-heavy interfaces. Those with anxiety disorders may need reassurance and clear explanations of what information is being shared and with whom. Credential wallets and platforms that ignore these needs create unnecessary barriers for populations that represent significant portions of the potential holder base.

The persistent failure to prioritize accessibility in credentialing systems reflects deeper assumptions about who deserves access to opportunity.

When public dollars fund credentialing infrastructure that excludes people with disabilities, limited English proficiency, or inadequate connectivity, these systems violate not only technical standards but fundamental principles of equity. Accessibility is not a technical checkbox or optional enhancement, it is the measure of whether credentialing systems genuinely serve all learners and workers, or merely those who already possess the abilities, resources, and circumstances that designers assume as default.

MARGINALIZED POPULATIONS: FROM GATEKEEPERS TO NEW BARRIERS

Verifiable digital credentialing systems are celebrated for their potential to democratize opportunity by possibly making skills more visible, credentials more portable, and pathways more accessible than they have been. Yet this vision falters when the systems themselves exclude the very populations they claim to serve. For many people from marginalized communities, Learning and Work Records (LWRs) issued as Verifiable Digital Credentials (VDCs) function not as liberating tools but as new gatekeepers, built without regard for their realities, resources, or rights.

The exclusions documented here are neither accidental nor technical oversights, they represent structural failures emerging from design processes that center privileged populations while treating marginalized communities as afterthoughts. From language barriers and device incompatibility to assumptions about identity verification and connectivity, current platforms reproduce existing inequities rather than disrupting them.

People with Disabilities: Despite representing 15% of the global population, people with disabilities encounter platforms designed without consideration for their needs. As previously shared blind and low-vision holders find wallet applications lack consistent screen reader compatibility, making basic navigation impossible. Images in credentials lack alternative text. Color-coded information excludes people with colorblindness.

Neurodivergent individuals face distinct barriers. People with ADHD struggle with multi-step processes lacking progress indicators. Those with autism encounter interfaces relying on ambiguous language and unstated conventions. People with dyslexia need specific fonts and spacing, accommodations platforms rarely provide. Those with anxiety require clear privacy explanations, typically obscured behind technical jargon.

Cognitive accessibility extends to people with intellectual disabilities, traumatic brain injuries, and age-related changes. Yet platforms design as if all holders possess identical cognitive capabilities, systematically excluding anyone whose profile differs from the narrow band designers consider 'normal.'

Immigrants and Refugees: Language barriers represent one of the most preventable forms of exclusion. Yet, most credential wallets and platforms operate exclusively in English, immediately excluding tens of millions with limited English proficiency. When translation exists, it covers only high-resource languages while ignoring hundreds of others. Automated translations often produce incomprehensible technical language.

The problem extends beyond interface text. Systems embed Western academic assumptions about how learning should be structured and documented. Informal learning, apprenticeships, family knowledge transmission, community skills and stories, becomes invisible. Multilingual proficiency has no space for documentation. Alternative literacies like oral traditions or practical demonstration lack representation in systems privileging written institutional validation.

International credentials pose additional obstacles. Three-year bachelor's degrees from India, German polytechnic certificates, or Vietnamese vocational training lack direct U.S. analogues that systems recognize. Refugees often possess no documentation, having fled circumstances where retrieving credentials was impossible. These populations find years of experience effectively erased.

Justice-Impacted Individuals: Returning citizens face stark digital exclusion. Despite being targeted in reentry pilots, justice-impacted individuals often lack smartphones, having been disconnected during incarceration. Those with devices rely on limited data plans or borrowed equipment unsuitable for credential management.

The digital literacy gap is also significant. Incarceration interrupts skill development during rapid technological evolution. Educational programs in facilities rarely include meaningful digital literacy instruction. Identity verification creates further barriers, such as platforms that require government IDs, current addresses, and digital footprints that returning citizens don't have. When systems prove too complex, individuals become dependent on intermediaries, undermining the autonomy and privacy digital credentials should enable.

Gig and Low-Wage Workers: Low-wage workers use older smartphones that struggle with contemporary applications. Platforms requiring recent operating systems or significant storage exclude those unable to afford upgrades. Data limitations create constant friction, such as workers with restrictive prepaid plans must make strategic choices about application use. Wallets or platforms requiring persistent connectivity become prohibitively expensive.

Shared device usage introduces complications systems don't anticipate. Families sharing smartphones cannot maintain separate wallets or accounts. Workers using public library computers encounter systems designed for personal devices. The assumption of universal individual device ownership pervades design, rendering systems inaccessible to those navigating technology through communal arrangements.

Other Systematically Excluded Populations: Beyond these groups, credentialing systems systematically exclude older adults navigating unfamiliar technology and experiencing age-related cognitive or sensory changes, rural residents with limited broadband infrastructure and facing geographic isolation from support services, unhoused individuals without stable addresses for account verification or secure device storage, Indigenous communities whose knowledge traditions and cultural practices don't map to Western credential frameworks, LGBTQ+ individuals whose identity documents may not reflect their authentic names or whose employment histories reveal discrimination, people with limited formal education whose extensive practical skills remain unrecognized, veterans whose military training fails to translate into civilian credentials, single parents and caregivers whose time poverty and caregiving skills go unacknowledged, individuals experiencing domestic violence who need privacy protections that systems don't provide, and people with mental health conditions facing stigma and capability fluctuations that rigid systems cannot accommodate.

Each population encounters distinct barriers, yet all share the common experience of systems designed without their participation or consideration.

The Structural Nature of Exclusion

These patterns share common roots: *design processes failing to center marginalized populations as full participants.* When developers work in resource-rich environments using current devices and reliable connectivity, serving populations like themselves, resulting systems encode their creators' circumstances. What developers experience as 'basic' functionality may represent insurmountable barriers for others. In short, *'who we are is what and for who we build.'*

The rhetoric of 'designing for the average person' systematically privileges already-advantaged populations. The statistical average represents no one's experience while centering young, educated, digitally literate individuals without disabilities. Those deviating become 'edge cases' whose needs are deferred.

The irony is profound: *populations who could benefit most from improved credentialing, those excluded by traditional systems, find themselves shut out again.*

Centering Equity as Foundation

Addressing these exclusions requires fundamentally rethinking the relationship between credentialing systems, emerging digital credential innovations and people from marginalized populations. Equity cannot function as an add-on, it must serve as the foundational principle shaping design from inception. This means engaging marginalized communities as true co-designers and decision-makers with authority over systems intended to serve them, not merely as research subjects or 'users'.

Until credentialing ecosystems center those most impacted by systemic injustice as full partners in design and governance, the promise of digital credentials and emerging verifiable credentials will remain unrealized, concentrated among populations already closest to the industry, opportunity while those who could benefit most encounter yet another barrier in a long history of exclusion.

SYSTEMIC BARRIERS IN PLATFORM DESIGN

Current credentialing platforms and wallets reveal a troubling pattern: *they are built for institutional convenience rather than holder needs.* This design orientation creates concrete barriers that undermine adoption and perpetuate exclusion, even when technical architecture functions as intended.

Research consistently documents how platform design creates barriers. Digital Promise's 2022 community college study found learners confused by wallet applications, with many assuming credentials held no value simply because their experience of them was poor (Digital Promise, 2023). Jobs for the Future's 2024 survey revealed that while 65% of jobseekers expressed interest in digital credentials, only 24% had actually used them, suggesting significant friction between interest and practice (Jobs for the Future, 2024).

Accessibility audits paint a starker picture. WebAIM's 2021 evaluation found over 86% of top website homepages had detectable accessibility violations, averaging 51.4 errors per page (Center for Democracy & Technology, 2021). Common failures included keyboard navigation traps, missing alternative text, insufficient color contrast, and failed mobile responsiveness tests.

These failures stem from who develops platforms and their priorities. For example:

- **Venture-backed startups** optimize for rapid growth and institutional sales rather than inclusive design. Product roadmaps prioritize bulk issuance, administrative dashboards, and enterprise integration while treating holder experience as secondary. Research on use also focuses on early adopters rather than populations who would benefit most. Accessibility becomes a deferred 'nice-to-have' rather than core requirement.

- **University research labs** develop innovative architectures but lack dedicated UX design capacity, resources for extensive testing with people outside student populations, or accessibility expertise. Academic incentives also reward novel technical contributions over usability refinement, producing systems that function as proofs-of-concept but falter beyond educated early adopters.

- **Enterprise vendors** build for institutional buyers such as registrars, HR departments, workforce agencies, not workers and learners. Sales cycles demonstrate value to procurement officers, not credential holders. Features reflect those writing checks, compliance reporting, audit trails, enterprise integration, resulting in holder-facing interfaces receiving minimal investment.

This produces systems designed with institutions as primary interest and holders as passive recipients. Features being prioritized include those which support administrative convenience, streamline issuance, batch processing, verification APIs, analytics dashboards, while neglecting holder experiences of understanding, storing, and sharing credentials. The following common design choices reveal this orientation.

- **Onboarding flows** require complex account creation with email verification, password requirements, and multi-factor authentication before holders can even view a credential, and are optimized for institutional security rather than holder completion.

- **Credential displays** present technical metadata (issuance dates, cryptographic signatures) prominently while burying information holders need such as *what the credential represents, how it compares to alternatives, where it's valuable.*

- **Sharing mechanisms** assume technical literacy around 'generating verifiable presentations' without explaining what these accomplish or who gains access.

- **Help resources** provide detailed administrator documentation while offering holders only generic FAQs.

Replicating Institutional Hierarchies

This institutional-first approach replicates existing power hierarchies. Institutions retain authoritative awarder/issuer and verifier roles while individuals remain passive, receiving credentials they don't understand, storing them in systems they don't control or can't easily access, sharing through processes they can't customize. The rhetoric of 'learner ownership' or 'worker centered' and 'portable credentials' rings hollow when holders lack comprehension, agency, and control to meaningfully leverage or even use their credentials.

The pattern mirrors historical dynamics where institutions held power and individuals had limited recourse. Digital systems promised to shift this balance, enabling individuals to assemble diverse credentials, control their representation, and direct their own narratives. Yet when platforms prioritize institutional convenience over holder empowerment, they simply digitize existing hierarchies, creating systems that are technically decentralized but functionally reproduce the same power imbalances they claimed to disrupt.

INADEQUATE SUPPORT INFRASTRUCTURE

Beyond interface design, credentialing platforms and wallets consistently fail to provide adequate support infrastructure for credential holders navigating unfamiliar systems. While consumer technology platforms routinely offer real-time chat support, comprehensive help centers in multiple languages, video tutorials, and community forums, credential wallets and platforms lag significantly behind these standards.

Most credential platforms provide only basic FAQ documentation, typically available exclusively in English and written at technical literacy levels that assume substantial prior knowledge. Help resources focus on institutional administrator needs such as *how to issue credentials in bulk, configure verification workflows, manage organizational accounts,* while offering minimal guidance for holders attempting to understand what credentials represent, how to access them across devices, or how to share them effectively with employers or educational institutions.

The absence of real-time support channels is particularly problematic. When credential holders encounter technical issues such as failed verification, inability to access wallets, confusion about sharing mechanisms, they typically have no immediate recourse. Email-based support, when available, operates on timelines measured in days rather than the minutes or hours that job application deadlines demand. This creates situations where individuals miss opportunities because they cannot resolve technical problems quickly enough, despite possessing valid credentials.

Video tutorials and visual demonstrations that could bridge literacy gaps and language barriers remain rare in credential platforms. While Jobs for the Future's 2025 market scan notes 'innovations in wallet-enabled storage and multilingual capabilities' as emerging features, these remain exceptions rather than standards (Jobs for the Future, 2025). Alternative formats that could serve people with diverse learning needs include: audio walkthroughs, simplified visual guides, step-by-step interactive tutorials, all of which are largely absent from platforms that assume all holders learn effectively from dense text documentation.

Community-based support structures that could provide peer assistance and culturally relevant guidance are virtually nonexistent in digital credentialing ecosystems. Unlike established technology platforms where communities of learners and workers troubleshoot problems, share workflows, and create tutorials in multiple languages, credential holders navigate systems in isolation. The absence of community onboarding programs, peer mentorship models, or navigator services, all common in other contexts serving vulnerable populations, leaves holders from marginalized communities without the social support structures that could compensate for inadequate platform design.

This support gap disproportionately affects people from marginalized communities who face compounded barriers. The platform assumption that all holders possess sufficient digital literacy, language proficiency, and technical confidence to navigate systems independently through text-based FAQs reproduces the exclusions discussed throughout this chapter.

WHEN POOR DESIGN CARRIES ECONOMIC COSTS

Digital credential systems, especially emerging verifiable systems don't merely inconvenience holders through poor design, they also impose direct and indirect economic costs that disproportionately burden those least able to afford them. These costs transform what should be enablers of opportunity into financial barriers, compounding the exclusions already documented in this chapter.

Direct Platform Costs

Some credential platforms charge holders for basic functions that should be free. Reissuance fees penalize individuals who change devices, lose access to accounts, or need credentials in different formats, situations that arise frequently for populations using shared or unstable technology. Verification fees create paywalls between credential holders and the opportunities their credentials should unlock, forcing individuals to pay repeatedly to prove the same qualifications. As explored in Chapter 5, these charging models concentrate costs on individuals rather than the institutions benefiting from credentialing infrastructure, effectively taxing holders for participating in systems purportedly designed to serve them.

Data and Connectivity Costs

Poorly optimized platforms impose hidden costs on holders with limited data plans. Bloated wallet and platform applications requiring hundreds of megabytes for download and updates consume expensive prepaid data that low-income workers carefully ration. Credential platforms demanding persistent internet connectivity to access locally stored credentials force holders to maintain active data connections, draining both data allowances and battery life. Video-heavy help resources and high-resolution credential images increase data consumption further. For populations managing restrictive data plans or relying on public WiFi with time limits, these technical choices translate directly into dollars spent on data or opportunities missed when digital credentials cannot be accessed without connectivity.

Opportunity Costs

The most significant economic burden emerges from missed opportunities resulting from inaccessible or incomprehensible credentials. A job seeker who cannot access their digital credential during a time-sensitive application loses not just that position but the income it would have provided. A worker unable to share credentials with an employer seeking to verify qualifications for a promotion foregoes advancement and higher wages. A student who cannot demonstrate competencies through digital credentials because platforms are inaccessible may fail to gain admission or scholarships, affecting long-term educational and economic trajectories.

These opportunity costs are difficult to measure but potentially devastating. Unlike the frustration of a confusing interface or the annoyance of poor customer support, missing employment, licensure, benefits enrollment, or educational opportunities has tangible economic consequences that cascade through individuals' lives and communities. When credential platforms fail at critical moments, such as during application deadlines, at verification checkpoints, in conversations with employers they don't merely waste time; they foreclose economic possibilities for populations already facing constrained opportunities.

The distribution of these costs follows predictable patterns. Those with stable employment, reliable technology, and financial cushions experience poor credential design as inconvenience. Those living paycheck to paycheck, managing limited data plans, and navigating precarious economic circumstances experience the same design failures as economic harm, transforming tools meant to expand opportunity into mechanisms that reinforce existing patterns of exclusion and disadvantage.

TOWARD INCLUSIVE RECORD AND CREDENTIALING SYSTEMS

Inclusion cannot be retrofitted. It must serve as the foundational principle shaping record and credentialing systems from inception. This may include determining who participates in design, whose needs drive decisions, and how success is measured. Building truly human-centered systems requires moving beyond tokenistic consultation to genuine co-design with the populations and groups these systems claim to serve, particularly those historically excluded from both traditional credentialing and emerging digital alternatives.

In the table below, we offer some core principles for inclusive design to help guide developers, designers, and project managers who want to develop more inclusive systems and programs.

Principle	Key Requirements	Why It Matters
Center Those Most Excluded	▸ Compensate community members as co-designers ▸ Create accessible participation structures ▸ Ensure decision-making authority, not just advisory input ▸ Maintain ongoing relationships beyond initial design ▸ Include learners and workers from marginalized groups	Populations facing greatest barriers possess irreplaceable knowledge about what barriers exist and how systems must function. Their needs must shape fundamental architecture, not be treated as 'edge cases' addressed after core functionality.
Prioritize Plain Language and Clarity	▸ Use plain language: 'proof of your skills' not 'verifiable credential' ▸ Short sentences, active voice, concrete examples ▸ Visual reinforcement of key concepts ▸ Progressive disclosure of complexity ▸ Clear communication throughout all interfaces	Clear communication increases comprehension, reduces errors, and improves task completion across all populations, not just those with limited literacy. Complexity doesn't signal sophistication; it privileges insider knowledge over usability.

Ensure Universal Technical Accessibility	▸ Screen reader compatibility ▸ Keyboard navigation without traps ▸ Sufficient color contrast ▸ Compatibility with older devices and OS ▸ Offline functionality ▸ Optimized for limited data ▸ Simplified modes for cognitive needs ▸ Clear progress indicators	Systems must function across full spectrum of devices, assistive technologies, and connectivity conditions holders actually use. WCAG compliance is baseline; true accessibility extends to cognitive, sensory, and situational needs.
Provide Comprehensive Support Infrastructure	▸ Short video tutorials ▸ Multilingual help resources with cultural context ▸ Real-time chat or phone assistance ▸ Community forums for peer support ▸ Trusted navigator programs ▸ Address both technical 'how-to' and conceptual understanding	Documentation alone cannot bridge diverse needs. Support must help holders understand what credentials represent, their value, how to discuss with employers, and troubleshooting, reducing anxiety and building confidence.

| Enable Diverse Forms of Expression | ▸ Narrative descriptions of skills and development
▸ Multimedia evidence (audio, video, visual portfolios)
▸ Community endorsements
▸ Self-assessments and reflections
▸ Flexibility beyond structured data fields | Particularly important for immigrants with international credentials, workers with informal skills, Indigenous learners, and anyone whose capabilities resist standardization. Rigid templates erase diverse expertise and alternative pathways. |

Source: LWYL Studio (2025)

EMERGING MODELS AND PROMISING PRACTICES

While fully inclusive credentialing systems remain rare, several initiatives demonstrate promising approaches:

▸ **State-Level Digital ID Implementation:** Colorado's my-Colorado digital ID program exemplifies human-centered government technology. The platform underwent extensive testing with diverse populations across regions, age groups, and digital literacy levels before deployment. Interface language prioritizes clarity over technical precision, privacy controls are prominent and comprehensible, and the system accommodates people who need assistance. The result is adoption by over 1.8 million residents, approximately one-third of the state's population, demonstrating that accessible design can achieve scale (State of Colorado, 2023).

- **Incremental Credentialing Frameworks:** The *Credential As You Go* initiative develops stackable, modular credentials designed for comprehensibility and practical utility. Rather than requiring completion of entire degree programs, learners receive recognition for incremental achievements that have labor market value. Framework documentation emphasizes equity considerations, acknowledges diverse starting points and pathways, and centers learner agency in assembling credentials that reflect individual goals rather than institutional structures (Credential As You Go, 2023).

- **Community-Centered Recognition:** The Open Recognition movement, coordinated through the Open Recognition Alliance (OAR), emphasizes human-readable badges, self-issued recognition, and community validation alongside institutional credentials. This approach acknowledges that meaningful recognition often emerges from communities and peers rather than formal authorities, particularly for populations whose learning happens outside institutional contexts. Open Recognition principles prioritize holder control, diverse recognition sources, and transparency about what badges represent (Open Recognition Alliance, 2023).

- **Learning from Promising Case Studies:** In 2025 LWYL Studio in collaboration with the US Chamber of Commerce Foundation T3 Network co-authored nine case studies on emerging practices in the issuance of learning and work records that center marginalized communities and diverse practices of recognition. Each case study documents the strategies, technologies, and partnerships that made these pilots possible, offering practical insights into what it takes to design, deploy, and scale verifiable credentials and interoperable systems. From incarcerated individuals building digital skill records to healthcare employers and workers using verifiable credentials for background checks, the collection showcases both innovation and impact in action.

These examples share common characteristics: meaningful engagement with diverse populations during design, commitment to accessibility and plain language, recognition that credentials serve holders rather than institutions, and willingness to challenge conventional assumptions about what credentialing systems should look like and whom they should serve.

The Path Forward

Building inclusive credentialing systems requires more than adopting specific features or following design guidelines. It demands fundamental reorientation of who holds power in credentialing ecosystems, shifting from institutional control to holder agency, from expert-driven design to community co-creation, from standardization that serves system efficiency to flexibility that accommodates human diversity.

This transformation will not emerge from technical innovation alone. It requires policy interventions that make accessibility and inclusion mandatory rather than optional, funding structures that reward equity over rapid deployment, accountability mechanisms that center the experiences of marginalized populations, and cultural change within the communities developing digital credentialing infrastructure and processes.

Until credentialing systems truly center those most impacted by exclusion as co-designers, decision-makers, and measures of success, the promise of digital credentials and especially verifiable digital credentials to democratize opportunity will remain unrealized, accessible only to populations who already possess the resources, capabilities, and cultural familiarity that systems assume as baseline requirements.

CONCLUSION

Digital credentialing systems that exclude people through inaccessible design, inadequate support, and institutional-first priorities don't simply fall short of their potential, they actively perpetuate the inequalities they claim to address. When platforms assume technological resources, digital literacy, language proficiency, and cultural familiarity that marginalized populations lack, they create new gatekeepers as formidable as the traditional credentialing barriers they purport to replace. The promise of digital credentials and especially verifiable digital credentials to democratize opportunity rings hollow when the systems delivering them systematically exclude those who would benefit most.

Inclusive design is not an aesthetic preference, a compliance requirement, or a feature to add in later iterations. It represents a fundamental question of justice: *whether verifiable digital credentialing systems will genuinely expand access to opportunity or merely digitize existing hierarchies of privilege*. The answer lies not in technical sophistication or cryptographic protocols, but in whose needs shape system architecture, whose experiences define success, and who holds power in credentialing ecosystems.

A system's inclusivity is measured not by its capabilities for those already advantaged, but by its accessibility to those facing the greatest barriers. If emerging verifiable digital credentials are to fulfill their promise of giving people control over their learning and work records to unlock opportunities, they must be designed with and for populations historically locked out, immigrants and refugees, people with disabilities, justice-impacted individuals, low-wage workers, older adults, rural residents, and all those whom traditional credentialing has failed to serve.

Until these communities participate as co-designers and decision-makers rather than passive recipients, digital credentials will remain tools of exclusion dressed in the rhetoric of empowerment.

FURTHER READING

1. Bergson-Shilcock, A. (2020). *The new landscape of digital literacy: How workers' uneven digital skills impact economic mobility and what policy can do about it.* National Skills Coalition. https://nationalskillscoalition.org

2. Center for Democracy & Technology. (2021). *Centering disability in technology policy: Issue landscape and recommendations.* https://cdt.org/wp-content/uploads/2021/12/centering-disability-120821-1326-final.pdf

3. Credential As You Go. (2023). *Incremental credentials framework playbook.* https://credentialasyougo.org/playbooks/incremental-credentialing-framework

4. Credential Engine. (2022). *Counting U.S. postsecondary and secondary credentials (4th ed.).* https://credentialengine.org/all-resources/counting-credentials/

5. Gilbert, R. M. (2019). *Inclusive design for a digital world: Designing with accessibility in mind.* Apress.

6. Jobs for the Future. (2024). *Jobseekers want digital credentials for skill-sharing: Are employers ready?* https://jff.org/jobseekers-want-digital-credentials-for-skill-sharing-are-employers-ready

7. National Skills Coalition & Federal Reserve Bank of Atlanta. (2023). *Closing the digital skill divide: The payoff for workers, business, and the economy.* https://nationalskillscoalition.org/wp-content/uploads/2023/02/NSC-DigitalDivide_report_Feb2023.pdf

8. Open Recognition Alliance. (2023). *Open recognition toolkit. Badge Wiki.* https://badge.wiki/wiki/Open_Recognition_Toolkit

9. Page, K. L. and USCCF T3 Network (2025). *LER Toolkit: Case study Collection.* Retrieved January 5, 2026, from https://www.t3networkhub.org/ler-toolkit/case-studies

10. Pew Research Center. (2016). *Digital readiness gaps.* https://www.pewresearch.org/internet/2016/09/20/digital-readiness-gaps/

FURTHER READING

11. State of Colorado Governor's Office of Information Technology. (2023). myColorado digital ID user experience. https://mycolorado.state.co.us/

12. U.S. Web Design System. (2022). *Inclusive design patterns research report.* Digital.gov. https://accessibility.digital.gov/ux/inclusive-design/

13. W3C Web Accessibility Initiative. (2023). *Web Content Accessibility Guidelines (WCAG) 2.2.* https://www.w3.org/WAI/WCAG22/quickref/

CHAPTER

10

CREDENTIALS THAT MISREPRESENT AND TOKENSIZE HUMAN EXPERIENCE

Verifiable digital credentials promise to accurately describe, communicate, and verify a person's abilities, achievements, and readiness for work or learning. In theory, they offer portable proof of what someone knows and can do. In practice, however, digital credentials, especially when automated, decontextualized, or poorly designed, often misrepresent the very people they claim to serve.

This chapter examines how learning and work records issued as verifiable digital credentials reduce complex human experiences to reductive checkboxes and standardized data fields. We explore four critical failures: *credentials that tokenize lived experience into quantifiable metrics, quality control gaps that render badges meaningless, universal taxonomies that erase contextual meaning, and translation breakdowns that occur when credentials cross contexts, from carceral settings to civilian workplaces, from informal learning to formal employment, from one cultural framework to another.*

When credentials flatten human capability into data points, they don't just fail to represent people accurately, they actively harm those who most need recognition of their skills, experiences, and knowledge.

REDUCTION: FROM EXPERIENCE TO CHECKBOX

To represent someone fairly through a credential requires nuance, context, and depth. Yet verifiable digital credentials routinely compress learning, work, and lived experience into decontextualized data points:

- A 'Teamwork' badge awarded after a single online module
- A 'Workplace Readiness' credential based on automated assessment scores
- A three-star rating in 'Digital Literacy' generated by algorithm

These credentials appear objective and data-driven. In reality, they obscure critical questions: *What kind of teamwork, conflict resolution in crisis situations or collaboration on routine tasks? In what context, leading a community organizing effort or participating in a corporate training exercise? Over what duration, sustained practice across years or a one-time demonstration? With what outcomes and for whom?*

Consider a formerly incarcerated person who coordinated peer support networks, mediated conflicts, and trained others in a prison education program for five years. Their transcript shows a 'Leadership Skills' badge, identical to one earned by a college student who served as treasurer of a campus club for one semester. The credentials appear equivalent. The experiences could not be more different.

By collapsing multidimensional human experience into standardized fields and universal categories, verifiable digital credentials create the illusion of precision while systematically erasing the very context that gives skills meaning. The result is not neutral, it actively disadvantages those whose experiences fall outside dominant institutional frameworks..

TOKENIZING HUMAN EXPERIENCE

Tokenization reduces rich, contextual human narratives into discrete, machine-readable data fragments. In digital credentialing systems, this manifests as:

- ▸ Binary completion markers that ignore depth, quality, or struggle ('completed' vs. 'not completed')
- ▸ Decontextualized rubric scores that strip away the circumstances of learning
- ▸ Complex interpersonal capabilities flattened into generic 'soft skill' badges

The violence of tokenization becomes most visible when applied to people whose experiences resist institutional categorization. Consider these examples:

Case	Description
The formerly incarcerated paralegal	A woman who spent eight years providing legal assistance to other incarcerated people, researching case law, drafting appeals, and training peer advocates receives a 'Resilience' badge upon release. The credential acknowledges her persistence but erases her legal expertise, her teaching ability, and the sophisticated advocacy work she performed under severe constraints. Potential employers see a soft skill; they miss the hard-won professional competencies.

The immigrant caregiver entrepreneur	A woman who organized and managed a network of 15 family childcare providers, handling scheduling, quality standards, regulatory compliance, and conflict resolution, receives credentials only as 'Volunteer Coordinator.' The system recognizes her coordination but not her project management, business operations knowledge, or systems-level leadership. Her entrepreneurial innovation becomes administrative support.
The community organizer	A 19-year-old who spent two years leading a campaign to restore bus service to a redlined neighborhood, conducting door-to-door surveys of 500 households, building a coalition across three tenant associations, testifying before the transit authority, training residents in public comment strategies, and ultimately securing $2.3 million in restored service, receives a badge in 'Civic Engagement.' The credential acknowledges that she participated in her community. It does not acknowledge that she conducted systematic needs assessments, managed cross-organizational partnerships, navigated bureaucratic systems, developed political strategy, or mobilized collective action that delivered measurable infrastructure wins. Employers see youth volunteerism; they miss the campaign management, policy advocacy, and systems change leadership.

In each case, the credential system recognizes what it can easily categorize while systematically erasing what matters most: *the substantive skills, the contextual intelligence, the creative problem-solving performed under constraint.*

Tokenization is not a technical limitation, it is a power arrangement. When credentials reduce people to data fragments, they transfer authority from individuals who lived the experience to institutions that control how experience gets named, valued, and interpreted. The person who navigated structural barriers becomes legible only through categories designed by those who never faced those barriers.

THE CREDIBILITY GAP: WHAT DOES A BADGE REALLY MEAN?

Verifiable digital credentials proliferate without gatekeepers. The technology enables anyone with basic technical knowledge to issue credentials that *look* official, complete with cryptographic signatures, metadata schemas, and blockchain verification. But verification proves only that a credential was technically issued, not that it represents meaningful achievement.

The issuer problem: No authoritative registry governs who can issue verifiable digital credentials or what standards they must meet. A badge in 'Data Analysis' might come from MIT's professional education program, a for-profit bootcamp with a 60% dropout rate, an automated LinkedIn Learning module, or an entrepreneur's weekend course. All carry identical visual and technical markers of legitimacy. The credential itself provides no signal to distinguish rigorous demonstration or assessment from pay-to-play certification.

The assessment gap: Many digital credentials emerge disconnected from any validation of actual capability. *They certify completion, not competence.* A learner watches six videos about project management, clicks through a multiple-choice quiz, and receives a 'Project Management Professional' badge, despite never having managed a project. The credential exists; the skill may not. While technical verification can occur, there is nothing substantive skill validation to verify.

The standards void: Identical credential names mask radically different requirements. One institution's 'Critical Thinking' credential requires a 20-page analytical essay with multiple revisions and peer review. Another's requires selecting correct answers on a 10-question quiz. A third's is auto-awarded after watching a video. All produce credentials labeled 'Critical Thinking.' All are technically verifiable. None are comparable.

The result: a credential marketplace that privileges form over substance. As of 2025, Credential Engine documented over 1.85 million unique credentials in the United States, a 72% increase in just three years, with minimal transparency about quality standards, assessment rigor, or labor market value. The proliferation creates noise, not signal.

For learners and workers, this chaos has consequences. Employers confronted with dozens of unfamiliar credentials default to traditional proxies: institutional brand names, degree requirements, years of experience. The very populations that alternative credentials were meant to serve, those without traditional degrees, those learning outside formal institutions, those building skills through lived experience, find their credentials dismissed as unverifiable noise.

Quality control failure doesn't just undermine trust in the system. It actively reinforces the gatekeeping mechanisms that credentialing reform was supposed to dismantle..

THE TYRANNY OF UNIVERSAL TAXONOMIES

Credentialing platforms structure skill and competency data through standardized classification systems, fixed taxonomies designed to make human capability legible to machines and employers. Examples include:

- **O*NET (Occupational Information Network)** - The U.S. Department of Labor's occupational classification system with over 1,000 occupation categories

- **ESCO (European Skills, Competences, Qualifications and Occupations)** - The EU's multilingual classification framework

- **CTDL (Credential Transparency Description Language)** - Credential Engine's standardized vocabulary for describing credentials

These taxonomies promise universality: *one language to describe all work, all skills, all capability. In practice, they function as straightjackets.*

The obsolescence problem. Taxonomies ossify the labor market as it existed when they were built. O*NET's occupation codes reflect industrial-era job categories; they struggle to capture platform work, portfolio careers, or hybrid roles that blend technical and interpersonal expertise. By the time new codes get added, a process that can take years, the work has already evolved. Workers performing cutting-edge labor find themselves classified in obsolete categories or shoehorned into approximate matches that misrepresent their actual expertise.

The cultural hegemony problem. These taxonomies encode particular assumptions about what constitutes legitimate work, learning, and how skills develop. They privilege Western, white-collar, credential-proxied pathways. 'Project management' is recognized when performed in corporate settings with Gantt charts and status meetings. The same coordinating, resource-allocating, stakeholder-managing work performed by a mutual aid organizer or a street market coordinator remains invisible, not because the skills differ, but because the context doesn't match the taxonomy's implicit norms.

The rigidity problem. Taxonomies demand discrete categories. Real human capability resists categorization. A food delivery worker navigating gig platforms deploys sophisticated digital literacy (app interfaces, algorithmic management systems, GPS navigation), customer service expertise (communication under time pressure, conflict de-escalation, cultural adaptability), and logistical optimization (route planning, time management, multi-platform coordination). Taxonomy codes capture 'Transportation and Material Moving' or 'Food Service.' They erase the cognitive complexity, the technological fluency, the entrepreneurial problem-solving that make the work possible.

The formalization bias. Taxonomies recognize work that bears institutional markers. A street vendor who sources inventory, manages cash flow, negotiates with suppliers, markets to customers, adapts offerings based on demand patterns, and builds customer loyalty is performing sophisticated business operations. But taxonomies classify 'entrepreneurship' through proxies: registered businesses, venture funding, formal incorporation. The vendor's expertise becomes illegible. Their credentials, if they have any, classify them as 'retail sales' or 'self-employed', categories that erase the business acumen, market analysis, and financial management they practice daily.

The problem is not that taxonomies exist, some standardization enables portability and comparison. The problem is the presumption of universality, the illusion that one classification system can faithfully represent the full spectrum of human work and learning. When credentials get locked into these taxonomies, they inherit the taxonomies' biases, blindnesses, and structural violence. Those whose work falls outside normalized categories don't just get misclassified, they get erased.

CONTEXT COLLAPSE: WHEN CREDENTIALS CROSS BORDERS

The most violent credential failures occur when learning and work records cross contextual boundaries, from carceral settings to civilian workplaces, from military service to civilian employment, from informal economies to formal sectors. Credentials that represent genuine expertise become illegible, dismissed, or actively stigmatizing when interpreted through a different frame.

From prison to workplace. A person spends five years in a prison education program earning credentials in welding, HVAC repair, or peer counseling. They complete hundreds of hours of coursework, pass industry certification exams, and gain extensive hands-on experience maintaining prison infrastructure or supporting fellow incarcerated people through crisis and transition. The credentials are technically valid, issued by accredited providers, aligned with industry standards, verified through assessment.

Yet when presented to civilian employers, these credentials trigger suspicion rather than recognition. Hiring managers see the issuing institution, a correctional facility or reentry training provider, and make assumptions: *the training must be inferior, the standards must be lower, the person must be untrustworthy.* Never mind that the welding certification required the same exam as its community college equivalent. Never mind that the peer counseling work demanded extraordinary emotional intelligence, conflict resolution skills, and cultural competency under conditions of extreme constraint. The carceral context contaminates the credential and the validated skills become invisible.

Even when employers look past the stigma, the credentials themselves often fail to translate. A 'Certified Peer Support Specialist' credential earned in prison doesn't map cleanly to civilian mental health job postings, which require credentials from state-approved community-based programs. The incarcerated person's hundreds of hours supporting others through addiction recovery, family separation, and institutional violence becomes categorized as 'volunteer work' or dismissed as irrelevant experience. Different context, different taxonomy, different meaning.

From military to civilian. A logistics operations supervisor in the Army manages supply chains for units of 500+ personnel across multiple deployment zones. They coordinate transportation, track inventory worth millions of dollars, optimize distribution networks under resource constraints, train teams, troubleshoot critical failures, and maintain operations under extreme pressure. Their military transcript shows '92Y Unit Supply Specialist' and lists military-specific certifications.

Civilian HR systems screen them out. The job title doesn't match 'Supply Chain Manager' search terms. The military credentials aren't recognized by automated resume scanners trained on civilian education. The hiring manager, unfamiliar with military occupational specialties, sees a gap: *no bachelor's degree, no civilian experience, no recognizable brand names*. The veteran's decade of sophisticated logistics expertise becomes 'military background', acknowledged as valuable in the abstract, unemployable in practice.

The translation problem isn't technical, military-to-civilian credential crosswalks exist. The problem is contextual illegibility. Civilian employers don't know how to read military credentials, don't trust their rigor, and default to proxies they understand: *university degrees, corporate brand names, civilian job titles.* The credential system that promised portability fails at the moment of cross-context interpretation.

From informal to formal economies. A woman runs an catering business from her home for fifteen years. She develops recipes, sources ingredients, manages finances, builds a client base, navigates food safety regulations, handles marketing, and employs seasonal help. Her expertise is substantial, her business successful. She has no credentials, no culinary degree, no *ServSafe* certification, no business license, no tax records that capture the full scope of her work.

When she applies for a position as a kitchen manager or seeks a microloan to formalize her business, the credentials system has no category for her. She is 'self-taught,' which codes as 'unqualified.' Her fifteen years of business operations don't count as 'entrepreneurship' because she never incorporated, never raised capital, never had a Dun & Bradstreet number. She can earn credentials now, take the ServSafe exam, complete a business fundamentals course, but those entry-level credentials misrepresent her actual expertise. *They credential her as a beginner when she is a master practitioner.*

The failure is systemic, not individual. These aren't edge cases or unfortunate misunderstandings. They are predictable failures built into how credential systems interpret context, or fail to. Credentials promise to make skills portable across contexts, but portability assumes commensurability: *that a skill means the same thing regardless of where or how it was developed.*

This assumption is false.
Skills are always contextual.

Expertise developed under constraint looks different than expertise developed with resources. Learning inside oppressive systems produces different, often deeper, forms of resilience, creativity, and problem-solving than learning in traditional classrooms.

When credential systems ignore context, they don't create a level playing field. They systematically disadvantage those whose learning happened in stigmatized, informal, or non-institutional settings. The people who most need credentials to validate their expertise, those without traditional educational access, those building skills through lived experience, those navigating structural barriers, find their credentials dismissed, mistranslated, or contaminated by the very contexts that made their learning necessary.

BEHAVIORAL CREDENTIALING AS SOCIAL CONTROL

Some of the most insidious credentials claim to measure not what people know or can do, but who they are: *their attitudes, dispositions, and behavioral traits*. Platforms issue badges and micro-credentials for 'grit', 'adaptability,' 'growth mindset,' 'professionalism,' 'emotional intelligence,' and 'workplace readiness.' These credentials perform a double erasure: *they reduce complex social dynamics to individual character traits, and they disguise normative judgments as objective assessment.*

The culture problem. Behavioral credentials assume universal standards for acceptable conduct, emotion regulation, and social interaction. They do not acknowledge, and often cannot accommodate, cultural variation in communication styles, authority relationships, conflict resolution, or professional demeanor. A 'professionalism' rubric penalizes Black and Latinx youth for code-switching, expressive communication styles, or forms of assertiveness that white assessors read as 'aggressive' or 'unprofessional.' These aren't deficits in professionalism, they're cultural differences being pathologized through assessment. *The credential doesn't measure capability; it measures conformity to white, middle-class behavioral norms.*

The neurodivergence problem. 'Collaboration' credentials typically assess group participation, verbal contribution in meetings, responsiveness to social cues, and comfort with unstructured social interaction. Autistic individuals, people with social anxiety, or those who process information differently may excel at collaborative work, producing high-quality contributions, synthesizing others' ideas, solving complex problems in partnership, while struggling with the specific behaviors the rubric measures. *They are credentialed as poor collaborators not because they cannot work with others, but because they collaborate differently than the assessment expects.*

The trauma problem. 'Emotional regulation' and 'resilience' credentials claim to measure how people manage stress, respond to setbacks, and maintain composure under pressure. But these assessments rarely account for how trauma, chronic stress, or ongoing survival strategies shape emotional expression. A formerly incarcerated person who appears 'guarded' or 'resistant to feedback' may be deploying adaptive strategies learned in environments where vulnerability meant danger. A domestic violence survivor who seems 'overly emotional' or 'anxious' may be navigating triggered responses to authority figures. The credential reads adaptive survival mechanisms as behavioral deficits.

The subjectivity problem. Unlike technical skills that can be demonstrated through performance tasks, behavioral traits are assessed through observation, interpretation, and judgment. Supervisors, teachers, or automated systems rate whether someone demonstrates 'leadership,' 'initiative,' or 'positive attitude.' These assessments are saturated with bias. The same behavior such as speaking up in meetings, questioning assumptions, proposing alternatives, gets coded as 'leadership' in white men and 'difficult' in women of color. Assertiveness becomes aggression. Confidence becomes arrogance. Independence becomes non-collaborative. *The credential doesn't measure the behavior; it measures how the assessor interprets the behavior through their own cultural lens and biases.*

The algorithmic intensification. Some platforms now use artificial intelligence to assess behavioral credentials automatically, analyzing video interviews for facial expressions, voice patterns, and word choice to rate 'enthusiasm,' 'trustworthiness,' or 'cultural fit.' These systems encode existing biases at scale while adding a veneer of objectivity. An AI trained on data from existing employees (predominantly white, neurotypical, middle-class) learns to recognize 'professionalism' as whatever those employees do. It then penalizes deviation, which systematically disadvantages anyone who doesn't match the training data's demographics or behavioral norms.

The ideology problem. Behavioral credentialing isn't neutral assessment, it's social engineering disguised as skill development. When institutions credential 'grit' or 'growth mindset,' they individualize systemic problems. A student who 'lacks resilience' isn't failed by under-resourced schools, racist discipline policies, or economic precarity, they simply need more grit. A worker who 'struggles with adaptability' isn't responding rationally to unstable employment, inadequate training, or impossible workloads, they need to develop a growth mindset. Behavioral credentials pathologize reasonable responses to unreasonable conditions, then offer those pathologies as explanations for inequitable outcomes.

The fundamental problem: *behavioral credentials measure compliance with dominant cultural norms while claiming to measure universal human capacities*. They credential conformity, then call it competence. And because behavioral traits are assessed through interpretation rather than demonstration, these credentials systematically advantage those who already align with assessors' expectations, those who look, speak, and behave like the people setting the standards.

ALGORITHMIC AMPLIFICATION: WHEN MACHINES INTERPRET HUMAN CAPABILITY

The latest evolution in credentialing embeds artificial intelligence throughout the entire lifecycle: AI generates learning records, issues credentials, interprets their meaning, and makes consequential decisions about people's futures. These systems prioritize machine-readability above human legibility: *credentials must fit standardized data schemas, map to predetermined taxonomies, and render human capability as structured datasets that algorithms can parse, rank, and filter at scale.*

This architectural choice, designing credentials for machines first, humans second, doesn't just enable automation. It hardwires algorithmic misrepresentation into the infrastructure.

Automated screening systems. Major applicant tracking systems (ATS) now scan credentials to automatically filter candidates before any human sees their application. The AI looks for exact keyword matches, credential completeness, institutional brand recognition, and conformity to expected patterns. A candidate with a non-traditional pathway such as a community college transfer, skills-based bootcamp, credentials earned while working full-time, gets scored lower than one with a linear four-year degree from a recognized university, even when their actual capabilities are superior. *The algorithm mistakes conformity to expected patterns for competence.*

Credential quantity over relevance. AI ranking systems often count credentials rather than evaluate them. Research by Upturn found that algorithmic hiring tools prioritize credential accumulation such as more badges, more certificates, more completed modules, over the relevance, rigor, or labor market value of those credentials. A candidate with fifteen micro-credentials from low-barrier online platforms may rank higher than one with two substantive credentials requiring demonstrated mastery. The system rewards credential collecting, not capability building. This particularly harms workers who invested time in fewer, deeper learning experiences or who couldn't afford to accumulate multiple credentials.

Pattern matching as proxy for potential. AI systems trained on historical hiring data learn to recognize 'successful' candidates by identifying patterns in existing employees' credentials, education, and experience. Then they filter new candidates against those patterns. The result: *the system replicates past hiring decisions, which means it replicates past biases.* If a company historically hired software engineers with computer science degrees from research universities, the AI learns that this pattern predicts success, not because it measures capability, but because it matches the training data. Qualified candidates with coding bootcamp credentials, self-taught portfolios, or degrees from less prestigious institutions get systematically filtered out. *The algorithm doesn't evaluate their skills; it evaluates their proximity to historical hiring patterns.*

The erasure of context and constraint. AI credential interpretation systems reduce people to their credentials and nothing more. They cannot read context such as why someone took seven years to complete a degree (they were working full-time to support family), why there's a gap in their learning record (they were caregiving, incarcerated, or dealing with medical crisis), why their credentials come from non-traditional sources (they couldn't access traditional education). These systems certainly cannot interpret constraint as capability, they cannot recognize that completing any credential while navigating structural barriers demonstrates precisely the persistence, problem-solving, and resourcefulness that employers claim to value. The algorithm sees only absence: *missing credentials, incomplete records, deviation from expected timelines.*

The motivation and aspiration void. Credential-matching algorithms optimize for past performance patterns, not future potential. They cannot assess someone's motivation to grow, their career aspirations, their willingness to learn, or their alignment with an organization's mission. A person transitioning careers who lacks credentials in the target field but demonstrates extraordinary self-direction and learning capacity becomes invisible. A formerly incarcerated person with relevant credentials who is deeply motivated to rebuild their life gets filtered out because their record shows an employment gap. *The AI sees credentials; it cannot see people.*

Data harms at scale. When these systems make errors and they do, constantly, the errors replicate at massive scale. One biased algorithm can screen thousands of qualified candidates out of consideration before a human ever reviews them. One flawed credential-matching system can channel entire populations away from opportunities they're qualified for. And because these systems operate as black

WHEN CREDENTIALS CAUSE HARM

boxes, proprietary algorithms, undisclosed training data, opaque decision logic, the people harmed have no way to understand why they were rejected, no avenue to contest the decision, no mechanism to demonstrate their actual capabilities outside the algorithmic framework.

The 2018 Upturn study documented these failures in detail such as algorithmic hiring tools that prioritize credential quantity over quality, that fail to surface candidates with non-traditional records, that systematically disadvantage precisely the populations that alternative credentials were supposed to help. Five years later, the problems have only intensified as AI systems become more sophisticated, more pervasive, and more trusted by employers who mistake automation for objectivity.

The fundamental architecture is flawed. When we design credentials to be machine-readable first, we optimize for algorithmic interpretation at the expense of human understanding. We privilege data completeness over experiential richness, pattern matching over contextual evaluation, and statistical correlation over genuine capability assessment. The machines get better at reading credentials. The humans, such as the workers, learners, and job seekers whose lives these systems shape, become increasingly illegible.

CONSENT FICTION. WHO CONTROLS YOUR CREDENTIALS?

Verifiable digital credential vendors and providers often promise learner ownership and control: 'You own your credentials.' 'Take your learning with you.' The reality is darker. Once issued, people lose meaningful control over how credentials circulate, who interprets them, and whether they can contest inaccuracies.

Invisible circulation. Digital credentials often enable frictionless verification, often without your knowledge. You apply for a job; the employer's system automatically verifies all credentials in your wallet. A professional network harvests your credentials to feed recommendation algorithms. You may have consented once when accepting the credential, but not to each subsequent use, interpretation, or inference.

Verification without permission. The technical architecture enables anyone to check credential authenticity. A potential employer verifies your credentials before contacting you. A landlord checks your educational background. A government agency cross-references credentials against eligibility requirements. You discover these verifications only indirectly, through rejection or unexplained decisions, if at all.

Permanence without recourse. Credentials persist in wallets, accounts, registries, and blockchain ledgers long after the contexts that produced them. A credential from a mental health crisis you've recovered from. A behavioral assessment from a job you left five years ago. These follow you, shaping algorithmic interpretations and employer evaluations and often after they've ceased to represent your current reality.

You cannot simply delete them. Issuers control revocation, not holders, and blockchain credentials that promise 'immutability' as a feature become a liability: *misrepresentations are permanently, cryptographically enshrined in ledgers you cannot edit.*

Misrepresentation without remedy. When credentials misrepresent you, paths to correction are limited or nonexistent. You might contact issuers, if you can identify and reach them. You might ask platforms to remove credentials, if they offer that option. More often, you have no recourse. The credential exists, systems interpret it, decisions get made. You may never know which credential triggered which rejection.

Compounding harm. Credentials feed into other systems. Your learning record influences hiring algorithms, which shape employment history, which affects credit scores, which determine housing access. A single misrepresentative credential cascades through interconnected systems, each amplifying the initial error, creating durable disadvantage that becomes impossible to contest as it spreads.

The consent theater. Most systems require consent: *click to accept terms, check boxes, agree to privacy policies.* This is procedural compliance, not meaningful consent. The forms are lengthy, technical, and opaque. They grant platforms sweeping rights to use and monetize credential data while disclaiming liability for errors. You 'consent' because refusal means losing access to the credential, the platform, the opportunity. This isn't informed consent, it's coerced acquiescence.

The fundamental issue is credential systems are built on an ownership fiction.

You 'own' your credentials like you 'own' your social media data, that is you have limited, revocable access to representations of yourself that others control and weaponize. True ownership would mean the right to see all uses, approve each verification, correct misrepresentations, delete permanently, and contest algorithmic interpretations. Few systems offer any of these rights. Most offer none.

TOWARD CREDENTIALS THAT RESPECT COMPLEXITY

To move beyond misrepresentation, credentialing systems must fundamentally redesign around human complexity rather than machine convenience. Ways to redesign may include:

▶ **Include narrative alongside data.** This might mean including multimedia evidence, context, and story, not just structured fields. A credential should accommodate the 'how' and 'why' behind the achievement, not only the standardized 'what.'

▶ **Enable Holder control.** The system must have genuine editing, revocation, and consent rights built into the architecture. People must control what gets shared, with whom, for how long, and retain the ability to update or remove credentials as their lives evolve.

▶ **Support rich contextual metadata.** The system must embed where, how, and under what conditions credentials were earned. A welding certification earned in prison is not identical to one earned in community college, not because the skills differ, but because the contexts do. Systems must preserve rather than erase these distinctions.

▶ **Inclusive taxonomies.** We must expand definitions of work, learning, and contribution beyond formalized, credentialed, institutional pathways. This might look like recognizing care work, community organizing, self-directed learning, and informal entrepreneurship as legitimate forms of expertise.

▶ **Enable distributed recognition.** This looks like enabling peers, communities, and collectives to validate experiences alongside institutions. Credential authority should not flow only from formal organizations to isolated individuals, but also horizontally through networks of mutual recognition.

▶ **Algorithmic transparency and consent.** Any use of AI to interpret, rank, or filter credentials must be transparent, contestable, and subject to informed consent. Disclose training data, make decision logic auditable, and provide meaningful mechanisms to challenge algorithmic determinations. Automated interpretation should supplement, not replace, human judgment.

These principles don't eliminate all risk of misrepresentation, human interpretation always involves judgment, power, and potential error. But they shift the architecture from erasure toward complexity, from extraction toward consent, from automation toward accountability. They ask credentials to serve the people they represent, not just the systems that interpret them.

DESIGNING FOR RECOGNITION, NOT REDUCTION

The purpose of a credential should be to recognize human capability and experience, not reduce it to data points. To expand access, not encode stigma. To reflect lived complexity, not erase it for algorithmic convenience.

This requires shifting power: *from institutions that credential to individuals who are credentialed. From algorithms that filter to communities that recognize. From systems designed for institutional compliance to systems designed for human flourishing.*

Credentials should serve the people they represent, not the platforms that profit from them, not the employers who use them as gatekeeping proxies, not the systems that demand machine-readable uniformity. When we design credentials that respect context, honor complexity, and center consent, we create infrastructure for recognition rather than surveillance, for opportunity rather than sorting, for dignity rather than reduction.

The question is not whether credentials can perfectly represent human experience, they cannot. The question is whether they amplify or diminish the people they claim to serve. Whether they open doors or reinforce gatekeeping. Whether they recognize capability or credential conformity.

We have a choice about which credentialing future we build.

CONCLUSION

Verifiable digital credentials were promised as tools of equity, portable proof of capability that would democratize opportunity, bypass gatekeepers, and recognize learning wherever it occurs. Instead, they frequently function as instruments of erasure and control. They reduce complex human experience to standardized checkboxes. They amplify algorithmic bias at scale. They extract consent without offering genuine control. They privilege institutional validation over lived expertise. They credential conformity while claiming to measure competence.

The harms are not incidental. Misrepresentation is not a technical glitch to be debugged. It is structural violence built into systems that prioritize machine-readability over human dignity, data cleanliness over contextual truth, and institutional authority over distributed recognition. *The consequences are material: jobs denied, opportunities foreclosed, expertise dismissed, potential wasted.*

We cannot build equitable credentialing systems by optimizing the current architecture. We must reimagine it entirely: credentials that preserve context rather than erase it, systems that center consent rather than extract it, infra-structure that recognizes complexity rather than reduces it. This requires humility about what credentials can capture, transparency about how they are interpreted, and account-ability when they cause harm.

No digital credential, no matter how verifiable, how de-tailed, how technically sophisticated, can fully contain a human life. The question is whether our systems acknowl-edge this limitation with grace, or whether they continue to mistake data fragments for whole people, checkboxes for lived experience, and algorithmic sorting for justice.

The choice is ours. The stakes could not be higher.

FURTHER READING

1. Bogen, M., & Rieke, A. (2018). *Help wanted: An examination of hiring algorithms, equity, and bias.* Upturn. https://www.upturn.org/static/reports/2018/hiring-algorithms/files/Upturn%20--%20Help%20Wanted%20-%20An%20Exploration%20of%20Hiring%20Algorithms,%20Equity%20and%20Bias.pdf

2. Brown, P., Lauder, H., & Ashton, D. (2011). *The global auction: The broken promises of education, jobs, and incomes.* Oxford University Press.

3. Credential Engine. (2021). *Counting U.S. postsecondary and secondary credentials.* https://credentialengine.org/wp-content/uploads/2021/02/Counting-Credentials-2021.pdf

4. Eubanks, V. (2018). *Automating inequality: How high-tech tools profile, police, and punish the poor.* St. Martin's Press.

5. Foucault, M. (1977). *Discipline and punish: The birth of the prison* (A. Sheridan, Trans.). Vintage Books. (Original work published 1975)

6. Jobs for the Future. (2021). *Opportunity youth pre-apprenticeship framework: Component 3—Culmination in one or more industry-recognized credentials.* https://www.jff.org/opportunity-youth-pre-apprenticeship-framework-component-3-culmination-in-one-or-more-industry-recognized-credentials/

7. O'Neil, C. (2016). *Weapons of math destruction: How big data increases inequality and threatens democracy.* Crown.

8. Raghavan, M., Barocas, S., Kleinberg, J., & Levy, K. (2020). Mitigating bias in algorithmic hiring: Evaluating claims and practices. *Proceedings of the 2020 Conference on Fairness, Accountability, and Transparency,* 469–481. https://doi.org/10.1145/3351095.3372828

9. Reich, J., & Ruipérez-Valiente, J. A. (2019). The MOOC pivot. *Science, 363*(6423), 130–131. https://doi.org/10.1126/science.aav7958

FURTHER READING

10. Sanchez-Monedero, J., Dencik, L., & Edwards, L. (2020). What does it mean to 'solve' the problem of discrimination in hiring? Social, technical and legal perspectives from the UK on automated hiring systems. *Proceedings of the 2020 Conference on Fairness, Accountability, and Transparency*, 458–468. https://doi.org/10.1145/3351095.3372849

CHAPTER

11

FRAGILE GOVERNANCE AND REVOCATION RISKS

CHAPTER INSIGHTS

▸ Governance structures: Who controls issuance, modification, and revocation of credentials?

▸ Power asymmetries: How centralized authority creates risks for credential holders

▸ Technical vulnerabilities: Blacklisting mechanisms, re-identification attacks, and fraud vectors

▸ Accountability gaps: Dispute resolution frameworks and their limitations in decentralized systems

▸ Real-world failures: Case studies examining credential disputes and revocation abuse

I ssuing learning and work records (LWRs) as verifiable digital credentials (VDCs) promises a more secure, portable, and learner-controlled way to represent achievements. Proponents emphasize cryptographic integrity, decentralized architectures, and holder autonomy.

Yet beneath these technical assurances lies a challenge that undermines many implementations: governance.

The question is deceptively simple: *Who decides?* Who has the authority to issue credentials, revoke them, correct errors, or determine whether a credential remains valid? What recourse exists when that power is misused, whether through negligence, bias, or deliberate exclusion?

These are not merely administrative concerns. Governance structures determine whether credentialing systems empower or marginalize people. Poor governance creates vectors for discrimination, enables arbitrary revocation, and opens pathways for surveillance and re-identification, even in systems designed to protect privacy.

This chapter examines the fragility of governance in digital credentialing systems, and emerging verifiable ones. We analyze the mechanisms of revocation and dispute

resolution, explore the risks of blacklisting and fraud, and present case studies that reveal how power asymmetries play out in practice. The goal is not to dismiss the potential of VDCs, but to illuminate the ethical and operational blind spots that persist, and to argue that without accountable governance, technical sophistication alone cannot deliver on the promise of equitable, trustworthy credentials.

WHO DECIDES WHAT CAN BE ISSUED, EDITED, OR REVOKED?

The rhetoric surrounding decentralized identifiers (DIDs) and verifiable digital credentials (VDCs) emphasizes individual agency and holder control. The reality is more constrained.

In most implementations, power remains concentrated in familiar hierarchies:

▶ **Issuers control** what credentials are created and when they are revoked

▶ **Verifiers determine** which credentials, and which issuers, they will trust

▶ **Holders are granted custody** of their credentials but excluded from governance decisions that affect their validity and use

This architecture replicates traditional credentialing power dynamics under a veneer of decentralization. The technologies may be new, but the structures of authority are not.

Unanswered Governance Questions

Despite growing adoption of VDCs, fundamental governance questions remain unresolved, or are answered differently across implementations with little transparency such as:

- **Issuer authority:** What qualifies an institution or individual to issue credentials? Who grants that authority, and under what criteria?

- **Registry maintenance:** Who controls the trust registries or accreditation lists that verifiers rely on? How are those lists updated or contested?

- **Error correction:** What process exists for appealing or correcting inaccurate credentials? How long does it take, and who bears the burden of proof?

- **Deletion rights:** Can credential holders request that credentials be suppressed, withdrawn, or permanently deleted? Under what conditions?

- **Revocation transparency:** When credentials are revoked, are holders notified? Are they given reasons? Do they have recourse?

The Limits of Holder Control

In practice, credential holders have limited agency. They cannot edit, annotate, or delete credentials without institutional approval. Revocation decisions are typically unilateral and opaque: *issuers can invalidate credentials without notice, explanation, or appeal.* Even in systems designed for portability, holders often discover their credentials have been revoked only when a verifier rejects them.

This asymmetry matters. When disputes arise, such as over grades, certifications, or employment records, holders find themselves at the mercy of gatekeepers who control both the credential and the process for contesting it.

The promise of self-sovereignty collides with the persistence of institutional control.

THE RISKS OF BLOCKLISTING, RE-IDENTIFICATION, AND FRAUD

Blocklisting and Silent Revocation

When verifiable digital credentials are revoked without explanation or warning, the effect resembles *digital block-listing*. The digital credential remains visible in the holder's wallet but is now flagged as *invalid, revoked, or unverifiable*. In some systems, issuers can and many do revoke credentials silently, in this the holder receives no notification and discovers the revocation only when a verifier rejects it

This opacity creates serious vulnerabilities:

- **Career disruption**: A professional license or certification is suddenly invalidated, preventing employment without the holder's knowledge until they attempt to use it.

- **Discriminatory signaling**: Revoked credentials can function as markers, flagging individuals with past legal issues, disciplinary actions, or contested employment histories in ways that perpetuate exclusion.

- **Retaliatory abuse**: Institutions can weaponize revocation authority, punishing whistleblowers, critics, or those who leave on contentious terms by invalidating their credentials.

Unlike traditional paper credentials that simply exist or don't, verifiable digital credentials can be retrospectively invalidated across entire networks. The holder's record is not merely questioned, it is actively negated, often with no accompanying due process.

Re-Identification Through Credential Patterns

Even when digital credentials are designed to be pseudonymous or privacy-preserving, metadata and usage patterns can re-identify individuals. The very features that make credentials verifiable, their cryptographic signatures, timestamps, and issuer identifiers, can become vectors for surveillance.

Research has documented several re-identification pathways:

- **Unique credential combinations**: In specialized fields or small communities, a specific set of credentials can uniquely identify an individual. A combination of a particular graduate degree, professional certification, and employer credential may apply to only one person in a region or industry.

- **Verification logs**: Each time a digital credential is verified, metadata is generated, who verified it, when, and potentially where. Aggregated verification logs can reveal job search patterns, geographic movement, or institutional relationships the holder may wish to keep private.

- **Blockchain transparency paradox**: Public blockchain-based revocation lists, while offering transparency and tamper-resistance, can expose patterns of exclusion. Observers can track which credentials are revoked, when, and potentially infer reasons based on timing or institutional context.

The Mozilla Foundation research has warned that decentralized identity systems can inadvertently leak identifiable patterns, particularly in low-population or high-sensitivity contexts where anonymity is most critical (Mozilla Foundation, 2022). The promise of privacy-preserving credentials collides with the reality of correlation attacks, a method of re-identifying individuals by analyzing and connecting seemingly unrelated pieces of information across different data sources or interactions. Even systems designed with privacy features (zero-knowledge proofs, selective disclosure, pseudonyms) can leak information through:

- ▸ Transaction patterns on blockchain
- ▸ Timing of credential issuance or revocation
- ▸ Which verifiers are checking which types of credentials
- ▸ Metadata that accumulates across multiple interactions

The more digital credentials someone holds and uses, the more correlation opportunities exist, creating a privacy paradox where active participation increases exposure.

FRAUD AND CREDENTIAL CONFUSION

The push toward open issuance and interoperability introduces new fraud vectors. As verifiable digital credential ecosystems become more permissive, distinguishing authentic credentials from fraudulent or misleading ones becomes harder, particularly when verification practices are inconsistent across platforms.

Key vulnerabilities include:

- ▸ **Self-asserted or issued credentials**: Some platforms allow individuals to issue credentials to themselves, which may be visually indistinguishable from institutionally issued ones. Without clear labeling or verification mechanisms, these can be mistaken for institutionally-issued credentials.
- ▸ **Weak revocation mechanisms**: Not all systems implement robust cryptographic revocation. Some rely on expired timestamps or manual checking rather than real-time revocation lists, making it difficult to detect invalid credentials that remain in circulation.
- ▸ **Spoofing and phishing**: QR codes and web-based verification portals can be replicated. Fraudulent credentials

with seemingly legitimate QR codes can link to fake verification pages designed to appear authentic, exploiting peoples' trust in visual verification cues.

▶ **Issuer impersonation**: In open ecosystems, nothing prevents bad actors from claiming to represent legitimate institutions. Without centralized vetting or trust registries, verifiers may struggle to distinguish real issuers from imposters.

The result is a verification crisis: *the more open and interoperable the system, the harder it becomes to establish trust without reintroducing centralized gatekeepers, the very entities these systems were designed to displace.*

DISPUTE RESOLUTION: WHAT HAPPENS WHEN SOMETHING GOES WRONG?

A central flaw in many verifiable and digital credentialing ecosystems is the *absence of meaningful dispute resolution*. When credentials contain errors, misrepresentations, or become contested, holders often discover they have no clear path to correction or appeal.

Unresolved Questions of Accountability

Consider these scenarios, each representing real risks within current systems:

▶ **Erroneous issuance**: A university issues a digital credential with the wrong degree title, graduation date, or honors designation. The holder discovers the error months later when applying for jobs. Who is responsible for correction? How long does it take?

▶ **Harmful or outdated language**: A digital credential uses terminology that has become discriminatory, or includes disciplinary language that mischaracterizes events. Can the holder request revision? Can they add context or rebuttal?

- ▶ **Issuer dissolution**: An employer goes bankrupt, a training provider closes, or a certification body loses accreditation. What happens to the digital credentials they issued? Who maintains verification infrastructure? Who handles disputes about those credentials?

- ▶ **Contested revocation**: An issuer revokes a digital credential claiming the holder violated terms of service. The holder disputes this characterization. Who arbitrates? What evidence is required? What are the timelines?

- ▶ **Cross-system conflicts**: A holder presents a digital credential that one verifier accepts and another rejects, citing different trust frameworks. Who resolves the discrepancy?

The Infrastructure That Doesn't Exist

Most digital credentialing systems and organizations lack the institutional scaffolding needed for accountability:

- ▶ **No formal grievance mechanisms**: Unlike traditional academic or professional credentialing bodies that have established appeals processes, many digital credential platforms often provide only customer service channels with no formal standing or protocols.

- ▶ **No independent adjudicators**: Disputes are typically resolved (if at all) by the issuing institution itself, the same entity whose decision is being contested. No neutral third party, ombudsperson, or arbitration board exists.

- ▶ **No transparent versioning**: Once a digital credential is issued, most systems have no mechanism for tracking edits, corrections, or disputes. If a digital credential is modified, the original version and the rationale for changes are rarely preserved or visible. In verifiable digital credentialing using cryptographic proofs the original credential is revoked and a new credential is issued.

▸ **No standardized timelines**: Even when dispute processes exist, they operate without clear timeframes. Holders may wait weeks or months for responses while digital credentials remain invalid or contested.

The Cost of Unresolved Disputes

Without clear protocols, errors and harms persist indefinitely. A misrepresented digital credential follows someone across job or education applications. A wrongly revoked certification blocks career advancement. A holder may spend months trying to correct a simple clerical error that, in a paper-based system, could be amended with a single letter from a registrar.

The consequences fall disproportionately on holders, who:

▸ Bear the burden of detecting errors

▸ Must navigate opaque or non-existent appeals processes

▸ Have no recourse when issuers are unresponsive or hostile

▸ Cannot compel corrections or force adjudication

Meanwhile, issuers face few consequences for poor governance. The asymmetry is stark: *those with the least power in the system, credential holders, carry nearly all the risk when something goes wrong.*

CASE STUDIES IN DISPUTE AND REVOCATION MISUSE

The following case studies illustrate how governance failures translate into real harm. Each demonstrates a different facet of the problem: *invisible revocation, unilateral power, the weaponization of credential systems, jurisdictional gaps, and economic coercion.*

Case Study: Healthcare Licensing: Revocation Without Notice or Appeal

During the COVID-19 pandemic surge in 2020, healthcare workers were rapidly credentialed for emergency response roles. Several states adopted digital credentialing systems to expedite verification and deployment.

What happened: Multiple nurses in one jurisdiction reported that their digital credentials for COVID-19 emergency training were suddenly revoked. The revocations were executed through an automated API call by the state licensing board's vendor system with no human review and no notification.

Impact: The nurses discovered the problem only when hospital systems rejected their credentials during onboarding for critical care positions. Some were turned away from shifts. Others faced delays of days or weeks while attempting to resolve the issue.

Resolution (or lack thereof): The licensing board eventually acknowledged that the revocations were erroneous, triggered by a database sync error that misidentified certain training completions. However, there was no formal appeals process. Nurses had to individually contact the board, provide documentation they had already submitted, and wait for manual reinstatement. No compensation or formal apology was issued.

Lesson: Automation without human oversight or appeal mechanisms can instantaneously invalidate digital credentials at the worst possible moment, during emergencies when workers are most needed.

FURTHER READING

1. Alexander, G. C., Tajanlangit, M., Heyward, J., Mansour, O., Qato, D. M., & Stafford, R. S. (2020). Use and content of primary care office-based vs telemedicine care visits during the COVID-19 pandemic in the US. *JAMA Network Open*, 3(10), e2021476. https://doi.org/10.1001/jamanetworkopen.2020.21476

2. Fraher, E. P., Pittman, P., Frogner, B. K., Spetz, J., Moore, J., Beck, A. J., Armstrong, D., & Buerhaus, P. I. (2020). Ensuring and sustaining a pandemic workforce. *New England Journal of Medicine*, 382(23), 2181-2183. https://doi.org/10.1056/NEJMp2006376

Case Study: Blockchain Credential Pilot: Silent Revocation and Job Denials

A Latin American university launched a blockchain-based credential pilot to issue tamper-proof student ID credentials. The system was promoted as giving students permanent, portable proof of enrollment and good standing.

What happened: When students transferred to other institutions or took leave, the university revoked their student ID credentials through the blockchain revocation registry. The credentials remained visible in students' digital wallets but were flagged as invalid on the blockchain. Students received no notification of the status change.

Impact: Several students discovered the revocations only when potential employers verified their credentials and found them invalid. In at least two documented cases, students lost job offers because employers interpreted the invalid credentials as evidence of academic misconduct or falsified records and not simply as evidence of non-enrollment.

Resolution (or lack thereof): The university maintained that revocation upon departure was standard policy. However, this policy was never communicated to students, nor was it explained that 'revoked' status would appear differently from 'graduated' or 'transferred' status. Students were told they could request letters of good standing, but these were paper-based and outside the digital ecosystem employers were using.

Lesson: Blockchain transparency does not equal accountability. Public revocation lists can stigmatize individuals without context, and 'tamper-proof' credentials become liabilities when revocation policies are opaque.

FURTHER READING

1. Grech, A., & Camilleri, A. F. (2017). Blockchain in education. *Publications Office of the European Union*. https://doi.org/10.2760/60649

2. Fedorova, E. P., & Skobleva, E. I. (2020). Application of blockchain technology in higher education. *European Journal of Contemporary Education, 9*(3), 552-571. https://doi.org/10.13187/ejced.2020.3.552

Case Study: Employer Microcredentials: Disciplinary Badges as Permanent Marks

A large retail corporation implemented an internal microcredential system to track employee training and skill development. Employees earned digital badges for completing modules on customer service, inventory management, and workplace safety.

What happened: Following a disciplinary hearing, one employee was issued a badge labeled 'Customer Disruption

Incident' that appeared in their internal credential profile alongside skill-based badges. The badge was issued by the HR department using the same system and visual format as training credentials.

Impact: The employee protested that the badge misrepresented the incident (a dispute with a manager, not a customer) and that displaying it as a 'credential' was punitive and humiliating. The badge appeared whenever managers reviewed the employee's profile for promotion or transfer opportunities. Colleagues could also view it through the company's internal credential-sharing feature.

Resolution (or lack thereof): The company had no formal process for contesting or removing the badge. HR representatives stated it was part of the employee's 'performance record' and would remain for two years. The employee filed a grievance through the union, but the case remained unresolved at the time of documentation. The badge system had been designed for skill recognition, not discipline, and no policies governed its misuse.

Lesson: Credentialing systems designed for recognition can also be weaponized for punishment. Without governance boundaries separating developmental records from disciplinary ones, employers can create permanent digital scarlet letters that follow workers throughout their tenure.

FURTHER READING

1. Ravenelle, A. J. (2019). *Hustle and gig: Struggling and surviving in the sharing economy.* University of California Press.

2. West, S. M., Whittaker, M., & Crawford, K. (2019). *Discriminating systems: Gender, race, and power in AI.* AI Now Institute.

Case Study: Cross-Border Recognition: When Interoperability Fails at the Border

A civil engineer trained in India held verifiable digital credentials including a bachelor's degree in civil engineering, a master's in structural engineering, and professional certifications from recognized Indian technical bodies. All credentials were issued as W3C-compliant verifiable credentials with cryptographic signatures. When the engineer migrated to Canada, they expected their credentials would be recognized under the country's stated commitment to digital credential interoperability.

What happened: The provincial engineering licensing board accepted digital credentials 'in principle' but maintained a closed trust registry of approved issuers, exclusively Canadian and select U.S. institutions. The engineer's credentials, while cryptographically valid and from nationally accredited Indian institutions, came from issuers not in the approved registry. The board refused to verify the credentials and provided no clear process for adding foreign issuers to the registry. When the engineer requested information about the criteria for registry inclusion, they were told it was determined by 'bilateral agreements' with no public timeline or application process.

Impact: Despite holding legitimate, verifiable digital credentials that met or exceeded Canadian engineering standards, the engineer could not practice or even sit for licensing exams. They were directed to pursue 'credential evaluation' through a third-party service at personal cost of several thousand dollars and 8-12 months of processing time, and potentially to re-take coursework at Canadian institutions, a process that could take 3-5 years and cost upward of $50,000. Meanwhile, their verifiable digital credentials sat unused in their digital wallet, technically 'valid' but functionally worthless.

Resolution (or lack thereof): After 18 months of advocacy through immigrant professional associations, the provincial board agreed to review credentials from the engineer's Indian university, yet only for that specific institution and only after a site visit and accreditation review that was funded by the applicant community. No systemic change occurred. Other foreign-trained engineers continue to face the same closed registry. The engineer eventually accepted work in an unrelated field at a fraction of their previous salary.

Lesson: Technical interoperability means nothing without governance interoperability. Trust registries can function as tools of professional protectionism, and 'global' credential systems remain fragmented by national gatekeeping. The promise of portable, borderless credentials collides with the reality of jurisdictional sovereignty and economic self-interest.

FURTHER READING

1. European Commission. (2020). Study on the movement of skilled migrants: Final report. Publications Office of the European Union, Luxembourg. Retrieved from https://op.europa.eu/

2. Gonzalez Gomez, M. (2024). Uncertainty and Colombian Immigrants' Encounters with the Foreign Credential Assessment System in London, Ontario.

3. Osaze, E. D. (2017). The non-recognition or devaluation of foreign professional immigrants credentials in Canada: The impact on the receiving country (Canada) and the immigrants.

4. Banjo, Y. M. (2025). Carriers of Coloured Credentials: The Devaluation of Foreign Credentials and Its Impact on the Career Trajectories of Professional African Immigrant Women in the Canadian Labour Market (Doctoral dissertation, University of Toronto (Canada)).

Case Study: Credential Held Hostage: Degrees as Collateral for Debt Collection

A public university in the United States adopted a comprehensive digital credentialing system, issuing verifiable diplomas, transcripts, and micro-credentials to all students. The university marketed the system as giving graduates 'permanent, portable proof of achievement' that they would control for life.

What happened: A graduate with $0 educational debt for college class and course fees, yet $280 in outstanding charges from unpaid parking tickets, a replacement ID card fee, and a disputed late library fine, had their digital credentials placed in administrative 'hold' status by the registrar's office. The credentials appeared normal in the graduate's digital wallet, complete with valid cryptographic signatures, but failed verification when employers or graduate schools attempted to validate them. The university's verification API returned a 'cannot verify, account hold' message. The graduate received no notification that their credentials had been put on hold or suspended.

Impact: The graduate lost two job offers after background checks revealed 'unverifiable' educational credentials. HR departments at both companies interpreted the failed verification as potential fraud. The graduate also missed the application deadline for a master's program after the admissions office rejected their transcript due to verification failure. Only after the third failed verification did the graduate contact the registrar and discover the hold. When they attempted to dispute the charges, particularly the $85 late library fine for a book they claimed to have returned, they were told the hold would remain until all charges were paid in full. No appeals process existed for disputed fees. The registrar's office stated that financial obligations must be satisfied before credentials can be released, treating the degree as institutional property rather than earned and paid for achievement.

Resolution (or lack thereof): Facing continued employment and educational barriers, the graduate borrowed money from family to pay the disputed charges. The hold was lifted within 48 hours of payment. However, the graduate had already lost both job opportunities and the graduate school application cycle. When they requested an explanation for why academic credentials could be suspended over non-academic debts, they were told it was 'standard university policy' dating back decades, now simply automated through the digital credentialing system. The university maintains this policy, and no governance review has occurred to separate academic standing (degree completion) from financial standing (disputed debts).

Lesson: When institutions retain indefinite control over credential validity, they can weaponize that power for non-academic purposes. Degrees become conditional rather than conferred, and credential systems become debt collection tools. The portability promised by digital credentials proves illusory when issuers can remotely invalidate them at will. What should be a one-time conferral of achievement becomes an ongoing relationship of power, where graduates remain perpetually subject to institutional authority.

FURTHER READING

1. Love, J. (2023). Stranded credits, stranded students: how students make sense of institutional debt policies and holds (Doctoral dissertation).
2. van Eck, M. (2025). Retaining Degree Certificates for Outstanding Student Debts: A Social-Justice and Ethical Issue in the Legal Profession. Obiter, 46(2), 293-311.

Patterns Across Cases

While each case study presents unique circumstances, from healthcare licensing errors to cross-border professional exclusion, they reveal consistent governance failures that transcend context, geography, and technology platform. These patterns are not accidental. They emerge from credentialing systems designed with institutional control and efficiency as primary values, relegating accountability, transparency, and holder rights to secondary concerns, or ignoring them entirely.

The table below synthesizes seven recurring failure modes observed across all five cases. Together, they illustrate how technical sophistication in credential systems can coexist with and even amplify fundamental injustices when governance is weak, opaque, or deliberately exclusionary.

Table. Recurring Patterns

Pattern	Description	Why It Matters
Invisible action	Revocations or modifications occur without notification to the holder	Holders discover credential invalidation only when denied opportunities, preventing timely response or correction
No appeal process	Holders have no formal mechanism to contest decisions	Errors, disputes, and abuse persist unchecked; power remains entirely with issuers
Asymmetric information	Issuers know what they've done; holders discover it only through denial or rejection	Creates fundamental unfairness where those affected are the last to know, preventing self-advocacy

Disproportionate consequences	Minor errors or policy changes result in job loss, delayed employment, or reputational harm	Small institutional mistakes or debts trigger catastrophic personal consequences with no proportionality principle
Power imbalance	Institutions face no accountability for errors or misuse; holders bear all costs	Risk and harm fall entirely on the most vulnerable parties while powerful actors remain insulated from consequences
Jurisdictional fragmentation	Technical standards don't guarantee recognition across borders or systems	'Interoperable' credentials fail at jurisdictional boundaries where governance, not technology, determines acceptance
Mission creep	Systems designed for verification are repurposed for control, punishment, or coercion	Tools built for empowerment become instruments of surveillance, debt collection, and institutional leverage

These are not merely technical failures or implementation oversights. They represent *governance failure by design*. The systems worked as intended, they simply intended the wrong things. They prioritized:

▶ Institutional control over individual rights

▶ Efficiency over fairness

▶ Automation over accountability

▶ Territorial interests over global mobility

▶ Convenience over human dignity

The promise of decentralized, holder-controlled credentials collides repeatedly with the persistent reality of institutional power. Without intentional governance design that centers accountability, transparency, and holder rights, digital credentialing systems, regardless of their technical sophistication, will replicate and potentially amplify the inequities of traditional credentialing infrastructures.

THE POLITICS OF TRUST REGISTRIES

Credential registries, which are curated lists of 'trusted issuers' or 'verified credentials' are often presented as neutral technical infrastructure, no different from DNS servers or certificate authorities. This framing obscures a fundamental reality: *every registry is a political artifact that encodes choices about whose knowledge counts, whose authority is recognized, and whose credentials matter.*

What Gets Excluded

In practice, trust registries reflect the priorities, biases, and economic interests of their stewards:

▸ **Community credentials disappear**: Registries exclude credentials issued by community organizations, worker cooperatives, mutual aid networks, or grassroots education initiatives, and not because these credentials lack rigor, but because issuers lack institutional status, or regulatory recognition or don't pay to play.

▸ **Peer learning becomes illegitimate**: Self-organized learning communities, open-source contributors, and peer-validated skills are systematically devalued when registries privilege only accredited institutions with legal standing

▸ **Commercial partnerships determine trust**: Some registries prioritize credentialing platforms and educational providers with commercial partnerships, integration agreements, or revenue-sharing deals, treating business relationships as proxies for issuer or credential quality

- **Geographic and linguistic bias**: Registries maintained by Global North institutions underrepresent issuers from the Global South, credentials in non-English languages, and culturally specific forms of knowledge and expertise

- **Legacy institutions maintain monopolies**: Traditional universities, professional associations, and government agencies are presumed trustworthy by default, while newer, more accessible, or more innovative credential issuers face opaque and often insurmountable barriers to inclusion

Opacity by Design

Most registries operate with little transparency about their governance and decision-making for what or who is included or excluded:

- **Inclusion criteria are vague or unpublished**: What qualifies an issuer as 'trusted' is often undefined, inconsistently applied, or known only to registry operators

- **Application processes don't exist**: Many registries offer no pathway for new issuers to apply, appeal rejections, or contest removals

- **Decisions are unilateral**: A small group of technical operators, standards bodies, or government agencies determines which credentials an entire ecosystem will recognize, with no participatory input from learners, workers, or credential holders

- **Removal is silent**: Issuers can be delisted without notification, explanation, or recourse, invalidating thousands of credentials instantly

Gatekeeping Legitimacy

Trust registries don't merely organize credentials, they also define legitimacy itself. When a credential isn't in the registry, it doesn't matter if it represents genuine learning, demonstrated competence, or years of experience. It is functionally invisible to verifiers who rely on the registry as their sole source of truth.

This creates a closed loop of power:

1. Registries define which issuers are trustworthy

2. Verifiers only accept credentials from registry-approved issuers

3. Issuers not in the registry cannot prove value to verifiers

4. Learners holding non-registry credentials are excluded from opportunities

5. Alternative forms of learning and credentialing are systematically delegitimized

Who Governs the Governors?

Because registry governance is rarely participatory, dominated instead by technical standards bodies, government agencies, or commercial platforms, it reinforces institutional dominance over learner agency.

The rhetoric of 'learner-controlled credentials' rings hollow when a handful of unaccountable gatekeepers determine which credentials any verifier will recognize.

The fundamental question remains unasked in most credentialing ecosystems: *Who decides who gets to decide?* Who has the authority to curate trust? By what process? With what accountability? And to whom?

Without clear answers, and without governance structures that include credential holders, people from marginalized communities, and diverse learning providers, trust registries become instruments of exclusion masquerading as neutral infrastructure.

THE ROLE OF VENDORS AND CLOSED APIs: PRIVATIZED GOVERNANCE BY DESIGN

Many digital credentialing systems are built by third-party commercial technology vendors who control the infrastructure for issuance, verification, and revocation. Through proprietary APIs and platform architectures, these vendors become de facto governors of credential validity and verification, often wielding more power over credentials than issuers or holders.

When credentials are revoked in vendor-operated systems:

- It typically happens through automated API calls that execute instantly.

- No human reviews the decision.

- No notification reaches the credential holder.

- The revocation might be triggered by an algorithm detecting policy violations, missed payments, or suspicious 'trust scores' based on opaque criteria.

- Holders discover the problem only when verifiers reject their credentials, often during critical job applications or enrollment processes.

- Appeals go to vendor customer service teams following policies designed to limit corporate liability, not ensure fairness.

- Meanwhile, issuing institutions have limited ability to intervene. They've outsourced the infrastructure, and with it, much of their authority.

This creates *privatized credential governance* where technology companies, not educational institutions or professional associations, make final determinations about validity, verification, and disputes.

Accountability becomes impossible to pin down. When things go wrong, is it the issuer's fault? The vendor's system? The algorithm? Distributed responsibility means no one is truly accountable.

Vendors operate under commercial law and terms of service, not educational, evaluation and learning data ethics. They have no fiduciary duty to credential holders. When their systems fail, remedy is limited to refunds or service credits, not restoration of lost jobs or educational opportunities. And once institutions adopt a vendor platform, switching becomes prohibitively difficult, giving vendors leverage to change policies unilaterally.

The fundamental question: *Should private companies control infrastructure that determines whose learning, experience and skills are recognized as legitimate?*

If credentials function as essential infrastructure for opportunity, their governance cannot rest on vendor discretion and non-negotiable terms of service. Without regulatory frameworks or transparency requirements, the ability to prove what you know and can do becomes subject to corporate policy rather than evaluationinformed by learning, recognition, and ethical data use protocols.

DESIGNING FOR ACCOUNTABILITY: WHAT CARE-CENTERED GOVERNANCE REQUIRES

Preventing the harms documented in this chapter requires more than technical patches or policy tweaks. It demands fundamental reorientation of how digital credentialing systems are designed, who controls them, and what values they encode.

Baseline Requirements for Accountable Systems

Any digital credentialing system affecting employment, education, or professional standing should meet minimum standards of fairness and transparency. Credential holders must be able to request corrections when credentials contain errors or outdated information, and to request deletion when credentials no longer serve their interests. Before any revocation takes effect, issuers must notify holders with clear explanation of what is changing, why, and what recourse exists, with sufficient time to respond before credentials become invalid.

When disputes arise, resolution cannot rest with the party whose decision is being contested. Systems need independent adjudication such as by neutral third parties with authority to investigate, reverse decisions, and compel corrections. For high-stakes credentials affecting livelihoods such as professional licenses, academic degrees, learning and work records, oversight bodies should include credential holders themselves, not just issuers and verifiers.

Transparency must be built into the architecture. Every digital credential action such as issuance, modification, revocation, should generate audit trails documenting who took the action, when, under what authority, and why. These logs should be accessible to credential holders and auditors, not buried in proprietary vendor databases.

Beyond Technical Solutions

Blockchain is not a substitute for good human and system governance. Immutable ledgers and cryptographic verification can enhance security, but they cannot ensure fairness, provide context, or restore opportunities lost to algorithmic errors. 'Trustless' systems still require human oversight, dispute mechanisms, and the ability to correct mistakes, capacities that pure technical solutions cannot provide.

The architecture matters less than the human and system governance wrapped around it. A centralized system with strong accountability mechanisms, transparent policies, and meaningful recourse *can be more trustworthy* than a decentralized system governed by opaque vendor terms of service.

The question is not whether the database is distributed, but whether power is.

Accountable digital credentialing requires recognizing that credentials are not neutral data. They are instruments of opportunity and exclusion. Systems that treat them as mere technical artifacts to be efficiently managed will continue producing the harms this chapter has documented.

Systems designed with credential holders' dignity and agency as primary constraints can point toward something better.

CONCLUSION

The architecture of trust in verifiable LWR and digital credentials depends on more than cryptographic signatures or interoperable metadata schemas. It depends on accountable, inclusive, and transparent governance structures.

It depends on people, their values, and their willingness to share power rather than merely automate its exercise.

The cases in this chapter demonstrate what happens when governance is treated as secondary to technology. Healthcare workers lose credentials to database errors with no appeal process. Students are silently blocklisted by blockchain registries without context or notification. Employees receive disciplinary badges that follow them indefinitely through systems designed for recognition. Qualified professionals are excluded by closed trust registries that protect territorial interests over human potential. Graduates have their degrees held hostage for minor debts, their achievements weaponized for debt collection.

These are not edge cases or implementation failures. They are predictable outcomes of systems designed without governance safeguards. Systems that prioritize institutional control over individual rights, efficiency over fairness, and automation over accountability.

The Choice Ahead

Without clear and fair processes for issuing, editing, disputing, and revoking credentials, and without meaningful recourse when systems fail and real consequences when institutions misuse their power, the promise of verifiable LWR and digital credentials will collapse into surveillance, exclusion, and error at scale.

Digital systems amplify both their benefits and their harms. The question is which we choose to scale.

Governance is not an afterthought to be addressed after technical standards are finalized and platforms are deployed. Governance is the central question. Who decides what credentials mean? Who has power to validate or invalidate them? Who bears the costs when systems fail? Who has voice in designing the rules? And who holds the powerful accountable when those rules are broken?

The promise of verifiable LWR and digital credentials, such as their portability, learner control, recognition of diverse learning, cannot be realized through technology alone. It requires political will to build systems where care, dignity, and accountability are foundational requirements, not aspirational additions. It requires recognizing that these records and credentials are not data to be managed efficiently, but records of human achievement that shape life possibilities. They deserve governance structures that reflect that gravity.

The technology exists. The standards are developing. What remains uncertain is whether we will build the governance frameworks necessary to ensure these systems serve human dignity rather than institutional convenience, and whether those most affected by digital record and credentialing decisions will have real power in shaping the rules that govern their lives.

FURTHER READING

1. Allen, C., et al. (2016). The path to self-sovereign identity. *Life with Alacrity blog*. Retrieved from http://www.lifewithalacrity.com/2016/04/the-path-to-self-soverereign-identity.html

2. Camenisch, J., & Lysyanskaya, A. (2001). An efficient system for non-transferable anonymous credentials with optional anonymity revocation. *Lecture Notes in Computer Science, 2045*, 93-118. https://doi.org/10.1007/3-540-44987-6_7

3. De Filippi, P., & Hassan, S. (2016). Blockchain technology as a regulatory technology: From code is law to law is code. *First Monday, 21*(12). https://doi.org/10.5210/fm.v21i12.7113

4. Eubanks, V. (2018). *Automating inequality: How high-tech tools profile, police, and punish the poor*. St. Martin's Press.

5. Gallagher, S. (2020, February 5). When ID systems fail: Lessons from India's Aadhaar. *MIT Technology Review*. Retrieved from https://www.technologyreview.com

6. Narayanan, A., & Felten, E. (2014). *Bitcoin and cryptocurrency technologies: A comprehensive introduction*. Princeton University Press.

7. Public pilot reports: Case documentation (2020–2025). Multiple jurisdictions and implementations.

8. Rannenberg, K., Camenisch, J., & Sabouri, A. (2015). *Attribute-based credentials for trust: Identity in the age of dynamism and privacy*. Springer.

9. Verhulst, S., Noveck, B. S., Caplan, R., Brown, K., & Paz, C. (2020). *Innovations in digital identity: How 13 countries are leveraging digital ID to advance economic and social goals*. Omidyar Network.

10. W3C Verifiable Credentials Working Group. (2022). *Verifiable credentials data model v1.1*. World Wide Web Consortium. Retrieved from https://www.w3.org/TR/vc-data-model/

12

FROM DESIGN TO HARM: CASE STUDIES

CHAPTER INSIGHTS

▶ *Health credentialing:* Distinguishing genuine opt-in from coercive pressure

▶ *Refugee employment pilots:* Portable credentials for displaced workers

▶ *Reentry credentialing:* Verifiable credentials in workforce reintegration programs

▶ *Skills-based hiring:* Bias amplification despite discrimination-reduction goals

▶ *Government authorization:* Implications of digitized governement authorization verification

Digital credentials and Learning and Work Records (LWRs) are frequently promoted as tools for opportunity and equity. Yet the lived experiences of those interacting with these systems reveal a more complex reality. Design choices, driven by institutional efficiency, innovation hype, or external funding priorities, can produce tangible harms, particularly for people navigating already-precarious systems.

This chapter examines five case studies where credentialing systems, despite well-intentioned goals, resulted in exclusion, coercion, stigma, or surveillance. These examples illustrate how misaligned design and governance can transform tools of recognition into instruments of risk. Each case reveals patterns that emerge when technical implementation outpaces consideration of power dynamics, context, and consequences for the most vulnerable groups of people.

The cases that follow span different contexts, from health credentialing and refugee employment to criminal justice reentry and automated hiring. While each operates in distinct domains, they share common fault lines: *the gap between voluntary participation and practical coercion, the tension between individual benefit and systemic*

*surveillance, and the ways algorithmic efficiency can am-
plify existing inequities.*

These are not cautionary tales about technology itself, but about the choices designers, implementers, and policy-makers make when deploying credential systems in environments marked by unequal power.

CASE STUDY: HEALTH CREDENTIALING AND THE PROBLEM OF COERCION

The Promise of Professional Recognition: In the wake of the COVID-19 pandemic, several state health departments launched digital credentialing pilots for frontline workers. One such effort, funded by a philanthropic foundation, offered behavioral health and care workers access to verifiable credentials for completing new training modules. The credential was framed as voluntary, an opportunity for professional recognition.

The Reality of Workplace Pressure: In practice, the system functioned differently. Workers reported pressure from supervisors to complete the training and claim the credential. In some settings, workers who declined were flagged as *non-compliant* during internal audits, despite the program's voluntary framing. The digital wallet required personal phone numbers as well as employer email addresses, effectively linking personal identity to workplace surveillance infrastructure.

Disproportionate Impact on Vulnerable Workers: The credential's impact fell disproportionately on women of color and immigrant workers in direct care roles. They reported confusion about their rights, loss of autonomy over their professional data, and fear of employer retaliation for non-participation. What began as a tool for professional recognition became a mechanism for compliance enforcement, one where the consequences of 'opting out' were never made explicit but were nevertheless real.

Design Failures and Governance Gaps: The design choices that enabled this coercion were structural, not accidental. The pilot's governance framework lacked clear boundaries between voluntary professional development and mandatory workplace requirements. No mechanisms existed to prevent employers from accessing participation data or using credential status in performance evaluations. The system's architects assumed individual agency in contexts where workers had little bargaining power, which is a fundamental misreading of the actual conditions in which direct care work occurs. When questioned about these dynamics, program administrators pointed to the *opt-in language* in their documentation, deflecting responsibility for how the system operated in practice.

Lessons for Equitable Credentialing: This case illuminates a broader pattern in digital credentialing: *the gap between designed intent and lived experience often reflects power imbalances that technology cannot resolve and may intensify.*

For credential systems to serve rather than exploit vulnerable workers, designers must account for the contexts of coercion that already exist in workplaces, particularly in sectors where workers have limited protections. Voluntary participation requires not just opt-in language, but structural safeguards against employer misuse, transparent governance that workers can influence, and accountability when systems are weaponized for compliance enforcement.

Harms experienced: The harms illustrated by this case include:

▸ the erosion of voluntary consent through structural pressure,

▸ expansion of employer surveillance through ostensibly beneficial tools, and

▸ the use of professional development systems to enforce compliance while maintaining the fiction of choice.

FURTHER READING

1. Aloisi, A., & De Stefano, V. (2022). Essential jobs, remote work and digital surveillance: Addressing the COVID-19 pandemic panopticon. *International Labour Review, 161*(2), 289–314. https://doi.org/10.1111/ilr.12219

2. European Centre for the Development of Vocational Training (Cedefop). (2022). Microcredentials for labour market education and training. *Cedefop Research Paper No. 87*. Publications Office. https://www.cedefop.europa.eu/en/publications/5574

3. Kato, S., Galán-Muros, V., & Weko, T. (2020). The emergence of alternative credentials. OECD Education Working Papers No. 216. https://doi.org/10.1787/b741f39e-en

CASE STUDY: REFUGEE EMPLOYMENT CREDENTIALS AND THE LIMITS OF TECHNICAL SOLUTIONS

The Promise of Portable Credentials: An international NGO piloted a blockchain-based credential wallet to support financial services and employment access for refugees resettled in two European countries. The system aimed to help displaced individuals prove their skills, education, and work history when traditional documentation had been lost or destroyed. The promise was compelling: *portable, verifiable digital credentials that refugees could present to potential employers across borders.*

Implementation Failures: In practice, the pilot encountered significant problems. Refugees were asked to enter sensitive personal data such as names, dates of birth, and education history into a wallet system they did not understand. The consent process, conducted through interpreters in crisis settings, provided little meaningful explanation of how the technology worked or who could access their data. Critical questions went unanswered: *Who owned the private keys? What happened if someone lost access? Could credentials be revoked? The system offered no assurance of data sovereignty.*

Lost in Translation: The technical infrastructure required standardization, which meant translating culturally diverse credentials into German or English taxonomies. A Syrian engineer's degree was mapped to European qualification frameworks that did not account for curriculum differences or institutional context. A teacher's decades of experience was reduced to checkbox categories that stripped away pedagogical approach and community knowledge. The result was credentials that looked standardized yet were deeply decontextualized.

Employer Rejection and False Promises: Employers struggled to interpret the digital credentials. Many questioned their legitimacy or could not assess equivalencies between unfamiliar educational systems and local standards. Most simply disregarded them, defaulting to traditional vetting processes or existing biases. The credential did not address structural barriers such as language discrimination, lack of networks, employers' risk aversion, or legal work restrictions. In many cases, it created false expectations of access while leaving fundamental obstacles unchanged.

Intermediary Control: NGO staff sometimes issued credentials without direct validation with refugees themselves. Case workers, operating under time pressure, made judgments about skill levels without deep knowledge of individuals' backgrounds. This concentrated power in intermediaries' hands rather than with credential holders. When refugees later discovered inaccuracies, correcting errors proved very difficult with an immutable blockchain system.

Harms experienced: Despite best intentions, harms were caused by the system as it:

▸ reduced complex refugee experiences to misinterpreted metadata,

▸ created false expectations of financial service and employment access, and

▸ shifted power toward technical and organizational intermediaries.

The pilot illustrated a broader pattern: *technical innovation cannot substitute for addressing power imbalances, cultural competence, or the structural barriers that exclude marginalized groups from opportunity.*

For credential systems to genuinely support refugees, design must begin with refugee leadership, accommodate multiple forms of community-based validation, and guarantee individual data control. Most importantly, credential initiatives must be embedded within broader cultural efforts to address employer bias and systemic devaluation of non-Western credentials such as recognizing that technology alone cannot solve fundamentally social and political problems.

FURTHER READING

1. Cheesman, M. (2022). Self-sovereignty for refugees? The contested horizons of digital identity. *Geopolitics, 27*(1), 134–159. https://doi.org/10.1080/14650045.2020.1823836

2. Madianou, M. (2019). Technocolonialism: Digital innovation and data practices in the humanitarian response to refugee crises. *Social Media + Society, 5*(3), 1–13. https://doi.org/10.1177/2056305119863146

3. Madianou, M. (2019). The biometric assemblage: Surveillance, experimentation, profit, and the measuring of refugee bodies. *Television & New Media, 20*(6), 581–599. https://doi.org/10.1177/1527476419857682

4. United Nations Development Programme. (2019, September 25). Turkey leads international public & private sector coalition to achieve SDGs by accelerating innovative solutions for refugees. https://www.undp.org/turkiye/press-releases/turkey-leads-international-public-private-sector-coalition-achieve-sdgs-accelerating-innovative-solutions-refugees

5. Ebert, J. (2018, November 29). Gravity Earth is using the blockchain to help refugees gain access to financial services [Interview]. *TechCrunch.* https://techcrunch.com/2018/11/29/gravity-earth/

CASE STUDY: CREDENTIALING IN REENTRY PROGRAMS

Portable Proof of Rehabilitation: A U.S. nonprofit launched a blockchain-based credentialing pilot for formerly incarcerated individuals completing job training programs. The goal was to create portable, verified credentials demonstrating skills and program completion to prospective employers.

Digital Barriers at Every Step: The credential wallet required email addresses, smartphone access, and multi-factor authentication. For individuals leaving incarceration, often without stable housing, reliable internet, or recent digital experience, these requirements created immediate exclusion. Digital literacy programs in prisons remain severely limited, leaving participants unprepared for smartphone-based systems and applicant tracking systems (ATS) now standard in employment applications.

One participant discovered their credential appeared on an open blockchain ledger with identifying information linked to their criminal history. Blockchain's immutability, which is designed to prevent tampering, conflicted with expungement laws and 'ban the box' policies meant to give formerly incarcerated individuals opportunities beyond their past.

Stigma Over Skills: When employers verified the credentials, many fixated on the incarceration context of learning rather than demonstrated competencies. The credential became a marker of criminal history rather than job readiness. Rather than opening doors, the system created a new mechanism for discrimination.

Harms Experienced: Participants experienced several harms such as:

- Digital exclusion where participants lacked smartphones, internet access, or digital literacy to manage authentication and credential claiming/sharing.

- Privacy violations: Blockchain's transparency created permanent digital records contradicting legal expungement rights and the principle that people can change.

- Amplified stigma: In some case the credentials reinforced rather than reduced employment discrimination, as employers used them to screen out applicants with justice involvement.

This case study demonstrates how credential systems designed without understanding the lived experiences of people from marginalized realities may work to amplify existing harms. Blockchain's core features, such as immutability and transparency, directly conflicted with reentry populations' needs for privacy and second chances.

A credential cannot overcome structural barriers such as employer bias, occupational licensing restrictions, housing instability, and the digital divide. Effective reentry support requires addressing these systemic issues, not technological solutions that embed disadvantage into their design.

FURTHER READING

1. Lageson, S. E. (2020). *Digital punishment: Privacy, stigma, and the harms of data-driven criminal justice.* Oxford University Press. https://doi.org/10.1093/oso/9780190872007.001.0001

2. Ogbonnaya-Ogburu, I. F., Toyama, K., & Dillahunt, T. R. (2019). Towards an effective digital literacy intervention to assist returning citizens with job search. *Proceedings of the 2019 CHI Conference on Human Factors in Computing Systems,* 1–12. https://doi.org/10.1145/3290605.3300315

3. Reisdorf, B. C., & Rikard, R. V. (2018). Digital rehabilitation: A model of reentry into the digital age. *American Behavioral Scientist, 62*(9), 1273–1290. https://doi.org/10.1177/0002764218773817

4. National Institute of Justice. (2022). Expungement: Criminal records as reentry barriers. *NIJ Journal, Issue 289.* https://nij.ojp.gov/topics/articles/expungement-criminal-records-reentry-barriers

CASE STUDY: SKILLS-BASED HIRING AND ALGORITHMIC BIAS

Reducing Degree Requirements to Expand Access: Multiple employers have piloted and some are adopting digital credential systems to support skills-based hiring. They are promised as a way to reduce reliance on four-year degrees and expand opportunities for candidates from non-degree backgrounds, including bootcamp graduates, certificate holders, and workers with demonstrated skills yet no formal degrees.

Algorithmic Screening Amplifies Bias: Several organizations have piloted third-party platforms that algorithmically score applicants based on digital credential metadata and skills tags, then automatically match candidates to job opportunities through filtering algorithms. Research has documented systemic patterns of bias in these AI-driven hiring systems showing that algorithmic hiring tools systematically disadvantage women, immigrant and Black applicants. For example, large language models favor white-associated names 85% of the time, while favoring Black-associated names only 9% of the time. Black male candidates face particularly severe discrimination, with models never favoring Black male-associated names over white male-associated names in comparative evaluations.

The Credential Equivalence Problem: These algorithmic systems further struggle to interpret equivalence across different learning settings. A project-based course in a coding bootcamp may develop identical competencies to a formal university course, yet the automated scoring systems lack frameworks to recognize this equivalence. Non-degree credentials such as peer recognition badges, community-based learning, or self-directed skill development, are often excluded from scoring entirely because they don't fit standardized taxonomies. This creates a new form of gatekeeping where the promise of 'skills-first' hiring actually reinforces traditional credential hierarchies.

Harms Experienced: Some of the harms experienced by participants in the pilots are documented as:

- Algorithmic exclusion, where women, Black applicants, and candidates from diverse backgrounds are systematically under-matched to opportunities compared to white men with equivalent demonstrated skills.

- Credential type bias where systems fail to recognize equivalence across learning contexts, privileging formal institutional-awarded credentials over equally valid non-degree and workbased pathways.

- Automation of discrimination in that what was marketed as an 'equalizing' filter becomes a new layer of bias, now with the veneer of objectivity that makes it harder to challenge than human decision-making.

Algorithmic hiring systems trained on historically biased data reproduce and amplify existing patterns of discrimination. When Amazon's AI recruiting tool was trained on a decade of predominantly male hires, it systematically downgraded resumes containing the word 'women.' When systems lack frameworks for recognizing non-degree credentials, they default to privileging traditional pathways, the very structures that skills-based hiring claims to disrupt.

The fundamental problem is not technological, it's structural. Algorithms trained on biased hiring histories will reproduce those biases unless organizations fundamentally redesign what data they use, how they define 'quality' across diverse credentials, and how they validate equivalence. Without addressing these root causes, skills-based and skills-first hiring risks becoming degree-based hiring with extra steps, now automated at scale and harder to contest.

FURTHER READING

1. Petersen, E., Denton, M., Calacci, D., & Bernstein, M. S. (2024). Gender, race, and intersectional bias in resume screening via language model retrieval [Working Paper]. Brookings Institution. https://www.brookings.edu/articles/gender-race-and-intersectional-bias-in-ai-resume-screening-via-language-model-retrieval/

2. Raghavan, M., Barocas, S., Kleinberg, J., & Levy, K. (2020). Mitigating bias in algorithmic hiring: Evaluating claims and practices. *Proceedings of the 2020 Conference on Fairness, Accountability, and Transparency*, 469–481. https://doi.org/10.1145/3351095.3372828

3. Sheard, M. (2025). Algorithm-facilitated discrimination: A socio-legal study of the use by employers of artificial intelligence hiring systems. *Journal of Law and Society, 52*(2), 373–398. https://doi.org/10.1111/jols.12535

4. Fuller, J. B., Sigelman, M., Barrero, J. M., & Sepulveda, F. (2024). *Skills-based hiring: The long road from pronouncements to practice.* Harvard Business School & Burning Glass Institute. https://www.hbs.edu/managing-the-future-of-work/Documents/research/Skills-Based%20Hiring.pdf

CASE STUDY: GOVERNMENT-ISSUED IDENTITY AND IMMIGRATION CREDENTIALS

Designing for Efficiency or Surviellance: As governments move to pilot and adopt verifiable digital credential formats for official documents, including visas, residency permits, and other immigration-related records, they and their vendors often promise greater efficiency, reduced fraud, and easier cross-border verification. However, these same systems can also introduce serious risks related to surveillance, control, and lack of individual agency, particularly for immigrants, asylum seekers, and non-citizens.

Uncovering Privacy and Civil Liberty Concerns: In recent pilots of mobile-based government immigration and identity credentials, digital verifiable credentials for work visas and digital drivers licences were issued via digital wallets that allowed individuals to store and present proof of legal status or authorization to work. While technically secure and cryptographically verifiable, several implementations raised privacy and civil liberties concerns. Specifically:

▶ *Verification logs:* When an individual presented their credential to an employer or border agent, the wallet system sent a request to the issuing authority to validate the credential, often logging device metadata, time, and location.

▶ *Lack of revocation control:* Credential holders had no ability to revoke or delete a credential, even if their legal status changed or they chose to stop using the system.

▶ *Unclear data sharing:* In some jurisdictions, verification logs were retained and accessible to government agencies, raising concerns about whether they could be used for immigration enforcement or profiling.

▶ *Revocation without recourse:* Reports from legal clinics noted that in some instances, changes in immigration policy led to credentials being remotely deactivated or invalidated, with no formal notice or appeals process for the individual.

These concerns have led civil rights organizations and immigrant advocacy groups to call for stronger protections, including selective disclosure, transparent data retention policies, and independent oversight of digital ID systems used for immigration and public services.

Rather than enhancing trust, the design of these systems, particularly those with verification processes that 'phone home' to the issuer, or transfer data to a verifier portal, can create conditions of persistent monitoring. For people from vulnerable populations, this turns a credential meant to increase access into one that risks exclusion and control.

FURTHER READING

1. Electronic Frontier Foundation. (2022, June 8). *Mobile driver's licenses are increasingly real—and they raise big privacy questions.* https://www.eff.org/deeplinks/2022/06/mobile-drivers-licenses-are-increasingly-real-and-they-raise-big-privacy-questions

2. World Privacy Forum. (2021, October). *The emerging digital identity ecosystem: Privacy, equity, and civil liberties implications.* https://www.worldprivacyforum.org/2021/10/digital-identity-ecosystem-privacy/

3. World Bank. (2018). *Principles on identification for sustainable development: Toward the digital age.* https://id4d.worldbank.org/principles

PATTERNS ACROSS CASE STUDIES

These case study examples differ in geography and population, yet they share core design and governance failures.

Consent as performance, not practice: Participation was framed as voluntary while being tied to benefits, employment, or basic services. Refugees who didn't register couldn't access food rations. Workers who declined credentialing were marked as uncooperative. When opting out carries penalties, consent becomes a formality.

True consent requires understanding, meaningful alternatives, and the ability to withdraw without punishment.

Credentials that encode more than they should: Systems designed to verify one thing such as your identity, completion, eligibility, are often expanded to track behavior, enforce compliance, or sort people into consequential categories. This scope creep rarely happens all at once. Each addition seems reasonable in isolation, but the cumulative effect is a credential that serves institutions rather than holders.

The permanence of digital credentials makes this worse: a paper certificate sits in a drawer, while a blockchain-based credential can be queried indefinitely.

Design assumptions that didn't travel: Many systems assumed stable internet, widespread smartphones, fixed addresses, consistent documentation, and trusted institutions. These assumptions fail predictably in rural communities, displaced populations, and contexts where state visibility is dangerous rather than empowering.

Effective design requires asking not just 'will this work?' but 'will this work for you, where you live, learn, and work?"

Accountability without recourse: When credentials contained errors or enabled harm, holders had few options. Appeals were slow, opaque, or nonexistent. This hits people from marginalized populations hardest as they are most likely to encounter errors and least equipped to fight them.

Accountability means more than a complaint form. It requires transparency, independent oversight, and genuine holder power.

CONCLUSION

Each case study began with legitimate goals: *improving access, increasing efficiency, enabling recognition.* The harms that followed weren't the result of malice but of blind spots, rushed timelines, and incentive structures that prioritized deployment over deliberation.

The pattern is consistent: *organizations design systems about people from marginalized communities rather than with them, optimize for institutional needs over individual agency, and treat deployment as the finish line rather than the starting point for accountability.*

Recognition is not control. A credential that opens doors for some while closing them for others or that trades visibility for vulnerability has failed its purpose. The technology isn't the problem. The problem is how it's governed, who shapes its design, and whose interests it serves.

Until those most marginalized, excluded, or diverse have genuine power to shape, refuse, and reform these systems, not as focus group participants but as codesigners and decision-makers, verifiable digital credentials will continue to replicate the inequalities they claim to address.

FURTHER READING

1. Ajunwa, I., Crawford, K., & Schultz, J. (2017). Limitless worker surveillance. *California Law Review, 105*(3), 735–776. https://doi.org/10.15779/Z38BR8MF4F

2. Allen, C. (2016, April 26). The path to self-sovereign identity. *Life With Alacrity.* https://www.lifewithalacrity.com/article/the-path-to-self-soveregin-identity/

3. Bogen, M., & Rieke, A. (2018). *Help wanted: An examination of hiring algorithms, equity, and bias.* Upturn. https://www.upturn.org/work/algorithmic-hiring/

4. Center for Democracy and Technology. (n.d.). *Digital identity and surveillance.* https://cdt.org/

5. Eubanks, V. (2018). *Automating inequality: How high-tech tools profile, police, and punish the poor.* St. Martin's Press.

6. Gilman, M. E. (2020). Poverty algorithms: A digital poverty agenda. *Stanford Technology Law Review, 23*(2), 236–268. https://law.stanford.edu/publications/no-23-2-2020/

7. Mozilla Foundation. (2020). *From openness to identity: A guide to addressing risks in digital ID.* https://blog.mozilla.org/netpolicy/files/2020/01/Mozilla-Digital-ID-White-Paper.pdf

8. Noble, S. U. (2018). *Algorithms of oppression: How search engines reinforce racism.* NYU Press.

9. Page, K. L. (2025). *Pilot and demonstration field notes, LWYL Studio (2020–2025)* [Unpublished raw data].

10. World Education Services. (n.d.). *Credential evaluation services for immigrants and refugees.* https://www.wes.org

13

PRIORITIZING TRUST, EQUITY, AND LIBERATION

▶ Human-centered and justice-first design alternatives

▶ Participatory codesign with marginalized communities

▶ Ethical data governance and community-controlled protocols

▶ Harm reduction principles for equitable technology

▶ Practical tools and implementation guides

What does it mean to build credentialing systems that actually serve the people they claim to recognize? If new technologies replicate the exclusions and harms of the systems they replace, innovation becomes entrenchment.

Too often, verifiable records and digital credentialing is framed as a neutral technical problem, a matter of standards, interoperability, and verification. But credentials are never neutral. They encode decisions about what counts as knowledge, who gets to define competence, and whose experiences are rendered visible or invisible. When these systems are designed without the people and communities most affected by them, they risk perpetuating the very inequities they promise to solve.

This chapter argues for a fundamentally different approach, one that centers the people and communities historically marginalized by credentialing gatekeepers. It draws on frameworks from design justice, data feminism, and community-based participatory research to reimagine how we build, govern, and sustain recognition systems. The goal is not to optimize existing structures, but to transform them.

What follows is a blueprint for that transformation, grounded in trust, equity, and collective liberation. It presents human-centered alternatives, ethical governance principles, participatory codesign methods, and practical tools for building systems that support dignity, not just compliance.

HUMAN-CENTERED AND JUSTICE-FIRST ALTERNATIVES

Digital credentialing initiatives often begin with technical architecture: *schemas, protocols, APIs*. But if these systems are to serve as tools of liberation, they must start with human needs and human relationships, particularly those among communities historically marginalized by education and employment gatekeepers.

A justice-first approach begins by asking:

▸ Who has been excluded or harmed by past credentialing systems?

▸ What would it look like to center their voices, needs, and leadership?

▸ What forms of recognition already exist in communities, and how might digital systems amplify rather than overwrite them?

Human-centered design does not mean simply making interfaces friendlier. It means designing *with*, not *for*, communities, as well as prioritizing agency, transparency, and context over efficiency and scale.

This reorientation requires more than good intentions. It demands structural change in how projects are funded, staffed, and evaluated. Funders must resource community engagement as a core part of infrastructure, not as an optional outreach. Design teams must include people with lived experience in decision-making roles, not just as interview subjects or focus group participants. And success must be measured not by adoption rates or technical milestones, but by whether the people most affected by credentialing systems report that their needs have been met, their dignity respected, and their power expanded.

BY CODESIGN WITH MARGINALIZED COMMUNITIES

By CoDesign is not stakeholder feedback. It is power-sharing. It recognizes that those closest to the problem hold essential insight into the solution. To codesign a credentialing system with community members and with justice in mind:

▶ **Engage community members** as paid, decision-making collaborators, not consultants brought in after key choices have been made.

▶ **Include people with lived experience** of systemic exclusion: *formerly incarcerated individuals, undocumented learners, caregivers, low-income workers.*

▶ **Schedule around their availability**, not institutional timelines.

▶ **Translate technical jargon** into shared, accessible and easy to understand language.

▶ **Validate existing informal, cultural, and peer-based recognition practices** rather than treating them as deficits to be corrected.

By CoDesign also means allowing for refusal. Communities must be free to say no, reshape the approach, or redefine what credentialing means in their context.

Participation without power is extraction.

Genuine codesign is slow, relational, and often uncomfortable for institutions accustomed to moving quickly. It requires sitting with ambiguity, revisiting assumptions, and accepting that community priorities may diverge from organizational goals. But this discomfort is generative. When communities shape the questions, not just react to proposed answers, the resulting systems are more likely to reflect real needs, earn trust, and endure beyond the initial funding cycle.

By Codesign is not a phase of the project. It is the project.

ETHICAL DATA GOVERNANCE AND PARTICIPATORY PROTOCOLS

Good technology without good governance is dangerous. Systems lacking ethical oversight can replicate harm at scale and faster and less visibly than their analog predecessors.

Ethical data governance must:

▶ **Define clear rules** about who can issue, edit, or revoke credentials

▶ **Require affirmative, ongoing consent** before data is shared

▶ **Provide holders with meaningful control** over their credentials, including deletion, annotation, and expiration settings

▶ **Include participatory governance structures** with community members, not just institutions or vendors

▶ **Ensure open documentation** and clear accountability pathways

Examples of ethical governance tools include:

▶ **Community charters** outlining platform usage, decision rights, and oversight mechanisms

▶ **Participatory audits** involving non-technical stakeholders in reviewing system behavior

▶ **Credential lifecycle maps** documenting data flow from issuance to revocation

Governance is not a back-end feature. It must be designed into every layer of the system and every ecosystem partnership.

Too often, governance is treated as a compliance exercise, that is something handled by legal teams after the product is built. But ethical governance must be embedded from the start, shaping technical architecture, partnership agreements, and human-facing interactions alike. This means creating accessible mechanisms for credential holders to understand how their data moves, to contest decisions made about them, and to participate in ongoing oversight. Governance bodies should include not only technologists and institutional representatives, but also community advocates, civil rights experts, and people who have been harmed by credentialing systems in the past.

Accountability without accessibility is theater.

DESIGN PRINCIPLES FOR HARM REDUCTION

Harm reduction is a practice rooted in public health and abolitionist movements. Applied to digital records and credentialing, it means recognizing that no system is neutral or harm-free, and that we must design to minimize risk, not just maximize efficiency.

Core principles:

▶ **Minimize data exposure.** Issue only what is necessary. Make revocation transparent. Use selective disclosure techniques such as BBS+ signatures or zero-knowledge proofs where possible.

▶ **Honor narrative complexity.** Include space for multimedia or story-based evidence alongside structured fields. Let holders explain credentials in their own terms, context matters.

- **Design for correction and contestability.** Build tools that allow holders to request edits, add context, or revoke credentials. Ensure deletion is genuine, not just visual suppression.

- **Build slow, build together.** Pilot with communities, not just institutions. Validate outcomes through lived experience, not adoption metrics. Resist the pressure to scale before trust is earned.

Harm reduction also requires humility. Designers and technologists must accept that they cannot anticipate every misuse, every unintended consequence, or every way a system might fail the people it was meant to serve. This means building in feedback loops, sunset clauses, and regular opportunities for reassessment. It means treating the first version of any system as provisional, subject to revision based on what communities report, not just what dashboards display.

A harm reduction mindset shifts the question from "How do we scale this?" to "How do we know if this is causing harm, and what will we do when it does?"

WHAT DOES LIBERATORY CREDENTIALING LOOK LIKE?

What might a record and digital credentialing system grounded in liberation actually look like?

Such systems would:

- **Validate** many forms of knowledge and skill, not just institutionally sanctioned ones

- **Be accountable to the people** they claim to serve, not only to the institutions, funders or vendor platforms

- ▸ **Make visible** what is often invisibilized: labor, care, cultural knowledge, resistance
- ▸ **Protect the right *not* to be recorded or credentialed**, to remain private, to opt out, to define success on one's own terms

These systems may look different from current platforms. They may be community-led rather than vendor-driven. They may be smaller in scale and slower to grow. They may prioritize relationships over reach.

That is not a limitation. That is the point.

Liberatory recordmaking and credentialing resists the logic that bigger is better and faster is smarter. *It asks us to measure success not by the number of credentials issued, but by the depth of trust built, the breadth of knowledge honored, and the degree to which power has shifted toward those who have historically been credentialed about rather than with.*

This vision will not emerge from a single platform or standard. It will emerge from many experiments, many communities, and many forms of recognition, some digital, some not. The work is not to build one system that serves everyone, but to cultivate an ecosystem where many forms of recognition can coexist, interoperate when useful, and remain accountable to the people they serve.

REAL-WORLD EXAMPLES AND INSPIRATION

The principles outlined in this chapter are not theoretical. Across sectors and geographies, organizations and communities are already building credentialing systems grounded in trust, equity, and self-determination. The following examples offer models, methods, and momentum for practitioners seeking to design differently.

Open Recognition Alliance is a global community advancing human-centered approaches to credentialing through open badges and peer-based recognition. Launched in 2016 the Alliance promotes practices that expand opportunities for individuals and communities to be recognized and to recognize others outside traditional institutional gatekeepers. The Alliance hosts the annual ePIC conference, which convenes practitioners, policymakers, and technologists working at the intersection of open education and recognition. Learn more: https://openrecognition.org

MyData Global is an award-winning international non-profit championing a human-centric approach to personal data. With over 100 organizational members and nearly 400 individual members across more than 40 countries, MyData advocates for individuals to have meaningful control over how their information is collected, stored, and shared. Their framework emphasizes affirmative consent, transparency, portability, and the right to access and correct one's own records. The MyData Declaration, which outlines core principles for restoring balance between individuals and organizations, provides a foundation for credentialing ecosystems that prioritize trust over extraction. Learn more: https://mydata.org

Murmurations Network offers a model for decentralized, interoperable data sharing that places control in the hands of people. Designed to support regenerative economy organizations and solidarity networks, Murmurations enables individuals and organizations to create and host their own profiles, which are then indexed and made available to aggregators building maps and directories. The protocol's core principles, data ownership, open sharing, and non-exploitation, demonstrate how distributed infrastructure can support collaborative mapping, trust networks, and community self-representation without relying on centralized databases. Learn more: https://murmurations. network

Reentry and Workforce Development Programs: Across the United States, reentry programs are exploring how credentialing can support returning citizens, individuals reentering communities after incarceration. According to the Prison Policy Initiative, formerly incarcerated people face an unemployment rate of 27.3%, nearly five times higher than the general population. Programs like STRIVE's Fresh Start in New York City offer workforce development, case management, and employment services using trauma-informed approaches. STRIVE reports that graduates of their previous reentry programs achieved a recidivism rate of less than 2%. These initiatives demonstrate that recognition and training programs designed with affected communities can become tools for economic restoration rather than further marginalization. Learn more: https://strive.org/freshstart

Digital Defense Playbook, developed by Our Data Bodies and distributed through Allied Media Projects, is a community-created toolkit for assessing risks and designing safer digital systems. Developed over three years through community gatherings, workshops, and interviews with residents in Charlotte, Detroit, and Los Angeles, the Playbook offers activities and frameworks for understanding how data-based technologies impact social justice work. It provides practical tools for communities to reclaim their data and build trusted models of safety, a model for how participatory research can inform technology design from the ground up. Learn more: https://www.odbproject.org

LWYL Studio and the By CoDesign Framework: LWYL Studio, founded by Dr. Kelly Page, is a social design studio that centers lived experiences in the design of learning and earning innovations. Through its *By CoDesign* framework, LWYL blends storytelling with consent-led research and community-based codesign to create inclusive, accessible solutions. Page's work on Inclusive Experience Design emphasizes that inclusive usability requires fundamentally rethinking who participates in design, whose knowledge counts, and how power is distributed throughout development process. The *By CoDesign* Book explores barriers to inclusive codesign, including power imbalances, resource constraints, and institutional resistance, and offers strategies for building systems accountable to the communities they serve. Learn more: http://lwylstudio.com

CONCLUSION

If we want to build systems that support liberation, we must start by caring.

- ▸ Caring who gets to define value.
- ▸ Caring who controls the data.
- ▸ Caring how harm is acknowledged and addressed.
- ▸ Caring whether people feel seen or surveilled.

Digital transformation is not inherently liberatory. Yet, it can be, if we are willing to slow down, share power, and design with the people who have the most to lose and the most to gain.

This is not just about records and credentials.
It is about justice.
The question is whether we will build accordingly.

FURTHER READING

1. Benjamin, R. (2019). *Race after technology: Abolitionist tools for the New Jim Code*. Polity Press.

2. Costanza-Chock, S. (2020). *Design justice: Community-led practices to build the worlds we need*. MIT Press. https://designjustice.mitpress.mit.edu/

3. D'Ignazio, C., & Klein, L. F. (2020). *Data feminism*. MIT Press. https://data-feminism.mitpress.mit.edu/

4. Eubanks, V. (2018). *Automating inequality: How high-tech tools profile, police, and punish the poor*. St. Martin's Press.

5. Lewis, T., Peña Gangadharan, S., Saba, M., & Petty, T. (2019). *Digital defense playbook: Community power tools for reclaiming data*. Our Data Bodies. https://www.odbproject.org/wp-content/uploads/2019/03/ODB_DDP_HighRes_Single.pdf

6. MyData Global. (2022). *What to make of data sovereignty*. https://mydata.org/2022/09/26/data-sovereignty/

7. Open Recognition Alliance. (n.d.). *Bologna Open Recognition Declaration*. https://openrecognition.org

8. Page, K. L. (2025.). *By CoDesign: Designing digital innovations better together*. LWYL Studio.

CHAPTER

14

WHO IS CREDENTIALING THE CREDENTIALERS? TRUST, OVERSIGHT, AND CONFLICTS OF INTEREST

► Conflicts of interest pervade credentialing evaluation when assessors profit from the systems they assess

► Nonprofit intermediaries increasingly blur the line between public advocacy and private revenue streams

► Independent, third-party oversight and community-led governance are not optional, they are essential

► Accountability structures must be designed into credentialing systems from the start, not retrofitted after harm occurs

D igital credentials and wallets are reshaping how we prove who we are and what we know. From block-chain-based academic records to employer-issued skill badges, these systems promise portability, verification, and trust. But a fundamental problem lurks beneath the surface: *who credentials the credentialers?*

When organizations that create credentialing standards also profit from consulting services to institutions adopting those standards, trust erodes. When vendors market 'independent' verification while selling the very systems being verified, conflicts of interest flourish. When nonprofit advocacy groups blur the line between public mission and private revenue, accountability suffers.

This chapter exposes governance gaps in verifiable record and digital credential ecosystems, where self-regulation masquerades as oversight, vendors assess their own products, and the entities writing the rules profit from them. We examine how current frameworks create perverse incentives, how conflicts of interest compromise independence, and why the absence of genuine third-party oversight threatens the integrity of the entire ecosystem.

More importantly, we explore pathways forward: *models for independent evaluation, community-led accountability mechanisms, and governance structures that separate standard-setters from those who profit from standards.*

Without rigorous checks and balances, verifiable record and digital credential systems will not merely fail to solve existing inequities, they will amplify them.

THE OVERSIGHT PROBLEM

In most verifiable record and digital credentialing ecosystems today, the organizations building, piloting, or selling systems are the same ones evaluating and reporting on them. The conflict of interest is not subtle, it is structural.

Consider the patterns:

- ▶ Vendors publish 'impact' reports without third-party validation
- ▶ Nonprofits partner with vendors for funding and co-promotion, then assess vendor products
- ▶ Government pilots measure success through grant-driven deliverables rather than actual utility for learners or workers

This arrangement constitutes what might be called an *evaluation-industrial complex: a network of assessments designed to confirm predetermined success rather than surface meaningful insight.*

The question is not whether bias exists, but how deeply it has been institutionalized.

WHO IS DOING THE EVALUATION NOW?

A review of current verifiable record, digital credentialing and wallet implementations reveals that evaluations are typically led by:

- **In-house teams at nonprofit intermediaries** who work within and directly benefit from the industry, including organizations such as Jobs for the Future (JFF), Digital Promise, Education Design Lab (EDL), and Credential Engine
- **For-profit vendors and consultants** hired by platform developers to assess developer products
- **Academic researchers** engaged through grant programs or vendor relationships that may create financial dependencies

In some cases, the evaluator is a formal partner of the project under assessment. In others, the evaluator and the pilot developer are funded by the same grant or foundation.

Independence between assessor and assessed is nominal at best, and often nonexistent.

THE PROBLEM WITH PARTNER-EVALUATORS

Evaluation reports produced under these conditions frequently lack the neutrality, transparency, and rigor that public trust requires. Common shortcomings include:

- No disclosure of financial or institutional relationships between evaluator and evaluated
- Selective success metrics that emphasize outputs (records/credentials issued) over outcomes (record/credentials used, understood, or valued)
- No data on harms, confusion, support needed, or system abandonment
- No community participation in evaluation design, data interpretation, or report review

The examples are not hypothetical. An annual evaluation and mapping of the LER Ecosystem by a vendor highlighted growing adoption rates by States and Institutions, yet but failed to report how many people actually accept, prefer, and use the records or later deleted the app or never understood what the record or credential represented. In another report, a microcredential platform claimed 'equity impact' without disaggregating data by race, income, or disability status.

These are not minor omissions, they are fundamental failures of evaluative integrity.

WHEN NONPROFITS ACT AS VENDORS

Nonprofit intermediaries play a valuable role in the verifiable record and digital credentialing ecosystem. They can convene stakeholders, translate research into practice, and advocate for policy as well as learner and worker interests. Yet, when these same organizations build tools, license products, offer pay-for-services, or receive significant funding from credential vendors, their capacity for independent evaluation becomes compromised.

The conflicts are widespread:

▸ **Credential Engine** promotes credential registry participation as a measure of transparency and issuer trustworthiness, yet does not validate the quality of credentials in its registry

▸ **Digital Credentials Consortium** positions itself as a research and standards body while developing technical infrastructure, receiving philanthropic funding to build and promote credentialing systems, and drawing its membership from institutions piloting those same systems

▸ **Digital Promise** operates a fee-based microcredential platform, charging per assessment, while publishing research promoting microcredentials as solutions for professional learning and workforce development. The organization both sells the product and produces the evidence base for its value.

- **Education Design Lab** consults with colleges on credential strategy while licensing its own Durable Skills Micro-credentials and charging per assessment. The organization advises institutions on what to build, and sells them the tools to build it.

- **1EdTech** publishes open standards freely, but the ability to influence their development is not open. Participation in technical working groups and steering committees requires paid membership, with contributing members gaining decision-making roles while lower-tier and non-members are excluded from the process.

- **Learning Economy Foundation** promotes open standards for global education transformation as a 501(c)(3) nonprofit, while its founding leadership simultaneously controls WeLibrary, LLC, the for-profit company that owns the LearnCard trademark and publishes the LearnCard app. The nonprofit advocates; the for-profit commercializes

These entanglements do not necessarily indicate bad faith yet when not transparent or declared they do erode public trust and raise urgent questions: *Who holds these systems accountable? And who is structurally positioned to do so?*

THE MYTH OF NEUTRAL INFRASTRUCTURE

Standards bodies and registries are often described as 'neutral infrastructure,' providing impartial foundations upon which others build. Yet neutrality is not a default condition. It must be earned through governance, transparency, and genuine independence. The problems we see emerge occur when:

- ▶ Standards are developed by consortia dominated by private companies with commercial interests
- ▶ Technical specifications are implemented without ethical review or impact assessment
- ▶ Interoperability rhetoric becomes a vehicle for vendor lock-in rather than genuine portability

Organizations like 1EdTech, W3C, and DIF are essential to protocol development. They create the technical scaffolding that makes verifiable records and digital credentials possible. But they do not provide ethical oversight, nor do they hold implementers accountable for how standards are deployed, or misused in the real world.

Technical neutrality is not the same as ethical neutrality, and infrastructure that shapes access to opportunity cannot be governed as though it were morally inert.

WE NEED INDEPENDENT OVERSIGHT

Genuine independent oversight requires more than a new committee or advisory board and we certainly don't need another nonprofit. Instead, it demands structural change: *mechanisms that are participatory, transparent, and rooted in public interest rather than industry convenience or vendor profit.*

We need:

Community Review Boards: Modeled on Institutional Review Boards in research ethics, community review boards would comprise people from directly affected groups: learners, workers, justice-involved individuals, immigrants, and others whose lives these systems shape. These boards would be empowered to review system goals, design choices, consent processes, and data practices before deployment, not after harm has occurred.

Public Technology Ombuds Programs: Independent offices, whether state-run, nonprofit-based, or federally chartered, would receive complaints and concerns from credential holders, trigger audits or investigations when patterns emerge, and publish transparent reports on misuse, harm, or exclusion. Unlike vendor complaint lines, these offices would have no financial relationship with the systems they oversee.

Open Evaluation Coalitions: Publicly funded consortia of evaluators with no vendor affiliations would conduct assessments using community-led framing of goals and metrics. Reports would be co-authored with participants, not delivered to them. Evaluation would become a form of collaboration, not extraction.

Credential Harms Registry: A shared public repository would document harms, grievances, and failures related to verifiable records and digital credentialing. Modeled on clinical trials registries or data breach disclosure databases, this registry would surface patterns, identify systemic risks, and create accountability that individual complaints cannot.

OVERSIGHT BY DESIGN, NOT JUST IN POLICY

Independent governance must be embedded in verifiable record and digital credentialing infrastructure from the beginning, not bolted on after deployment. Policy statements and ethical guidelines matter, but they are insufficient without technical and procedural mechanisms that make accountability operational.

Design-level accountability includes:

- **Open logs** of credential issuance, revocation, and disputes that can be audited by independent parties
- **Credential metadata** that documents the awarder and issuer's governance structures, oversight policies, and accountability mechanisms
- **Verifiable transparency layers** such as attestations of independent audit completion or community co-design participation
- **Holder-controlled settings** that give individuals genuine agency over credential visibility, expiration, portability, and sharing

Systems that claim to empower individuals while denying them meaningful control are not trustworthy, regardless of their technical sophistication.

FUNDING OVERSIGHT WITHOUT COMPROMISING INDEPENDENCE

One persistent barrier to independent evaluation is funding. Most credentialing ecosystems depend on grant money or vendor revenue, which makes evaluators financially dependent on the entities they assess. Breaking this dependency requires intentional structural separation.

Practical approaches include:

- **Separated budgets**: Funders should allocate development and evaluation funds to different organizations, with explicit prohibitions on cross-subsidy or coordination
- **Public funding for oversight**: Governments should fund third-party oversight bodies through public

agencies rather than through contractors with vendor relationships

▸ **Community benefit agreements**: Vendors seeking public contracts or foundation support could be required to co-fund independent oversight mechanisms governed by affected communities, not by the vendors themselves

Independence is not achieved by declaration. It is achieved by designing financial structures that make capture difficult and accountability possible.

CASE STUDY: PARTICIPATORY TECH OVERSIGHT IN ACTION

In 2022, a coalition of digital rights organizations piloted a participatory oversight framework during a *citywide smart infrastructure deployment*. Although the project involved urban technology rather than verifiable records or digital credentialing specifically, its governance model offers transferable lessons.

Key features included:

▸ **Community-led review** of system goals, anticipated risks, and acceptable trade-offs before deployment

▸ **Transparent harm logs** published regularly and accessible to the public

▸ **Mandatory data protection audits** conducted by parties with no financial stake in the system's success

▸ **Opt-out pathways and redress mechanisms** that gave affected residents meaningful choices and recourse

The model demonstrated that governance by the governed is not utopian. It is achievable when oversight structures are designed with community authority as a foundational requirement rather than an afterthought.

DESIGNING OVERSIGHT FOR WALLET BUILDERS

Digital wallets are not neutral containers. They are access points and their builders wield significant power over what verifiable record and digital credentials can be stored, verified, or shared; what activity logs are retained and by whom; and what interfaces reveal or obscure holder agency.

Oversight of wallet builders should include:

- **Usability audits** conducted with low-literacy and multilingual people to ensure interfaces are genuinely accessible
- **Accessibility standards enforcement** that goes beyond compliance checklists to test real-world functionality
- **Prohibitions on dark patterns** and behavioral nudges that manipulate people toward choices that benefit the platform rather than the holder
- **Transparency requirements** for data retention policies, API integrations, and third-party data sharing

Open-source wallets are not exempt from these requirements. Openness is a technical characteristic, not an ethical guarantee.

Being auditable is not the same as being audited, and being open is not a substitute for being accountable.

Who Should Not Lead Oversight?

To avoid performative governance, oversight that exists in name but not in function, clarity is needed about who is unsuitable for leadership roles in evaluation and accountability.

The following should not lead oversight efforts:

- **Vendors** evaluating their own products
- **Nonprofits** with product licensing or revenue-sharing arrangements
- **Consultancies** receiving implementation contracts from the systems under review
- **Foundations** using evaluation metrics to justify prior investment decisions

This does not mean these actors have no role to play. Vendors can provide technical documentation. Nonprofits can facilitate convenings. Consultancies can offer implementation expertise. Foundations can fund independent work. But none should serve as the primary author of evaluation reports or the ultimate arbiter of system accountability.

WHAT THE PUBLIC DESERVES

Issuing learning and work records as verifiable digital credentials touches fundamental rights: *the right to be recognized, to access opportunity, to tell one's own story as well as the right to be forgotten.*

Systems that exercise this kind of power must meet the highest ethical standards, not merely technical compliance or regulatory minimums. The public deserves:

- **Oversight bodies** grounded in human rights frameworks, not just efficiency metrics
- **Design reviews and compliance audits** led by affected communities, not conducted on their behalf by vendors
- **Accountability for harm caused**, and for harm ignored, minimized, or explained away

A verifiable record or digital credentialing system that cannot be questioned cannot be trusted. And a system that resists independent scrutiny has already answered the question of whose interests it serves.

CONCLUSION

The systems that record and credential us must themselves be credentialed, not by contracts, endorsements, or marketing claims, but by public trust.

Trust cannot be purchased, asserted, or encoded. It must be earned through transparency, maintained through accountability, and shared through governance structures that center the people these systems claim to serve.

The future of verifiable records and digital credentials does not rest solely in code or policy. It rests in governance, and in the willingness to ask uncomfortable questions: *Who does this system serve? Who decided? And who has the power to change it?*

FURTHER READING

1. Brown, S., Davidovic, J., & Hasan, A. (2021). The algorithm audit: Scoring the algorithms that score us. *Big Data & Society, 8*(1). https://doi.org/10.1177/2053951720983865

2. Krimsky, S. (2013). Do financial conflicts of interest bias research? An inquiry into the 'funding effect' hypothesis. *Science, Technology, & Human Values, 38*(4), 566–587. https://doi.org/10.1177/0162243912456271

3. Lam, K., Lange, B., Avraamidou, M., Hasan, A., & Floridi, L. (2024). A framework for assurance audits of algorithmic systems. In *Proceedings of the 2024 ACM Conference on Fairness, Accountability, and Transparency (FAccT '24)*. ACM. https://doi.org/10.1145/3630106.3658972

4. Moore, D. A., Tanlu, L., & Bazerman, M. H. (2010). Conflict of interest and the intrusion of bias. *Judgment and Decision Making, 5*(1), 37–53. https://doi.org/10.1017/S1930297500002023

5. Open Government Partnership. (2024, December 10). *Participatory approaches in digital governance: Five examples from the field.* https://www.opengovpartnership.org/stories/participatory-approaches-in-digital-governance-five-examples-from-the-field

CLOSING REMARKS: AS WE BUILD IT, WHO REALLY BENEFITS?

CHAPTER INSIGHTS

- ▸ Credentials reflect power. Design must center equity and consent.
- ▸ Intentions don't prevent harm, oversight and accountability do.
- ▸ Build with people, not just for scale or speed.

Credentials are never just about recognition. They are about power.

Learning and Work Records (LWRs), digital wallets, verifiable digital credentials (VDCs), and blockchain-based identity systems are not neutral infrastructure. They reflect design choices, governance decisions, funding incentives, and social values. They determine who gets seen, who gets included, who is trusted, and who gets left behind.

This book began with a question: *What if we cause harm?*

Through fifteen chapters, we've surfaced the many ways digital verifiable credentialing systems, when built without care, consent, or accountability, can become new tools of exclusion, surveillance, and reduction. These systems hold promise. Yet without deep and participatory oversight, they risk turning into yet another layer of bureaucratic control, one that makes people legible to systems before making them whole to themselves.

What does that mean? It means a person becomes a list of verified competencies before they can be understood as a complete human with context, relationships, and self-knowledge. They are reduced to what can be measured, not recognized for all they are.

WHAT WE'VE LEARNED

1. Credentialing Has Always Been Political

From medieval university diplomas to modern digital badges, credentials have always defined who counts and who decides. They carry with them histories of colonialism, patriarchy, and racism. They can open doors, but they can also close them.

Consider: literacy tests used as voting credentials to disenfranchise Black citizens. Professional licensing systems that excluded women from medicine and law. Immigration documents that determined human worth based on national origin. Credentials have been used to gatekeep, exclude, and surveil, often under the guise of standardization and quality assurance.

To build equitable systems today, we must begin with this historical truth. Digital credentials may feel new, but the power dynamics are ancient.

2. Intentions Aren't Enough

Many digital credentialing projects are funded by philanthropy, designed by idealistic teams, and built with equity in mind. Yet even the best intentions can replicate harm if power and governance remain unexamined. Without active participation from the people most impacted, systems inevitably reflect the priorities of institutions, not individuals.

Good intentions don't prevent harm. Oversight and accountability do.

3. Access and Usability Matter

If a worker can't understand or share their credential, it doesn't matter how technically compliant it is. If a returning citizen can't log in or is re-identified through metadata, the harm is not theoretical, it's real. If a credential requires a smartphone, stable internet, and English literacy, it serves those who already have access, not those who need it most.

Systems that fail to prioritize clarity, accessibility, and human experience are not broken. They are working as designed, for someone else.

4. Vendors Are Not Neutral Actors

Credentialing platforms, wallet providers, and technical intermediaries are often presented as infrastructure. Yet they are businesses, sometimes backed by venture capital, sometimes masquerading as nonprofits. Many are deeply entangled with the very systems they are meant to evaluate.

Consider this: when a workforce board hires a credentialing platform vendor who also serves as the 'independent' evaluator of credential quality, who wins? Not the workers. Not the learners. The vendor wins twice, once through the service contract, once through the evaluation that validates their own product.

This creates a dangerous concentration of power: vendors control data, define standards, and shape narratives, often without accountability.

5. Governance Is the Core Challenge

We have no shortage of standards, protocols, and pilots. What we lack is governance: structures that ensure decisions are made *with*, not just *for*, communities. Participatory governance, transparent dispute resolution, harm registries, and credential holder rights are not nice-to-haves. They are baseline requirements for trust.

Yes, verification matters. Fraud prevention is a legitimate concern. But verification designed solely to prevent fraud, without considering dignity, consent, and power, becomes surveillance.

We can build systems that are both trustworthy and humane. We simply haven't prioritized it.

FROM RECOGNITION TO LIBERATION

This is the shift we must make. We must move from credentialing systems that simply make people legible to institutions, to systems that make institutions accountable to people.

What does that look like in practice?

It means a community health worker who has trained informally for a decade can credential their own expertise and have it recognized, not because an institution approved it, but because their community vouches for it.

It means a formerly incarcerated person can control whether their prison education appears on their transcript, or can annotate it with context about what they learned and how they've grown.

It means an immigrant can hold credentials from their home country without being forced to translate, notarize, or validate them through systems that don't recognize their worth.

That means designing systems that:

- **Respect consent and control**, no credential moves without permission
- **Recognize informal, community-based, and lived knowledge**, not just institutional degrees
- **Allow individuals to annotate, revoke, or reject credentials**, including the ability to refuse participation entirely
- **Ensure privacy, safety, and recourse when harm occurs**, with real accountability mechanisms
- **Make institutions visible and answerable**, who issued this? Under what authority? With what biases?

It means credentialing with care, not just with code.

THE ROLE OF FUNDERS AND POLICYMAKERS

Philanthropic funders and state/federal agencies have a profound influence on how credentialing ecosystems evolve. Too often, they prioritize:

- Scalability over nuance
- Innovation over stability
- Pilot metrics over lived outcomes
- Vendor partnerships over community ownership

Funders must support:

- **Long-term community codesign**, not just consultation, but shared decision-making from the start
- **Independent evaluation**, by researchers and organizations with no financial ties to vendors

- ▶ **Oversight bodies unaffiliated with vendors**, governance structures with teeth
- ▶ **Research into unintended consequences and lived impact**, including harm tracking and public documentation

Public investments must be tied to public governance. If taxpayer dollars fund a credentialing system, the public, especially those most affected, must have a say in how it operates.

THE RESPONSIBILITY OF BUILDERS AND EVALUATORS

If you are a designer, developer, researcher, or evaluator in this space: you are responsible for what you build or assess. This responsibility includes:

- ▶ **Naming conflicts of interest**: who funds you? Who benefits from your work?
- ▶ **Rejecting extractive vendor partnerships**: even when they offer the most resources
- ▶ **Centering people, not platforms**: designing for human dignity, not technical elegance
- ▶ **Making room for refusal**: building opt-out mechanisms and honoring when people say no
- ▶ **Slowing down**: pilots that rush to scale often leave the most vulnerable behind

It is not enough to say 'this is open source' or 'this is standards-based.' Equity is not built by default. It is built by deliberate, often difficult, decisions, decisions that may cost you funding, partnerships, or speed.

Those are the right decisions.

BUILDING THE FUTURE WE NEED

The good news: alternatives are emerging.

▶ Community badging and peer-issued credentials that honor local knowledge

▶ Open-source wallets with built-in consent controls and multilingual interfaces

▶ Participatory governance charters that give credential holders real power

▶ Data cooperatives and worker-controlled credential platforms

▶ Multilingual, low-literacy, mobile-friendly tools designed with, not for, marginalized communities

▶ Harm registries and public accountability mechanisms that name when things go wrong

These are not fringe ideas. They are the future, if we choose to fund and build them. If we choose what is needed, what matters, and what centers impact over optics.

A NOTE ON HOPE

This book has focused on harm not to provoke despair, but to protect possibility.

Naming harm is an act of care. It says: *I see what could go wrong, and I refuse to let it happen to you.*

Hope is not blind optimism. It is the belief that we can build better if we slow down, listen, and co-create with the people whose lives these systems claim to serve.

We can credential care work, the labor that sustains families but appears nowhere on a resume. We can credential mutual aid organizing, the work that builds community power but earns no certificate. We can credential migration navigation, the multilingual, legally complex work that helps people survive but generates no badge. We can credential parenthood, translation, cultural labor, survival, and resilience, not just job titles and degrees.

We can validate community leadership. We can recognize lived experience. We can design credentialing systems that reflect not just what a system wants to see, but who a person truly is.

WHERE DO WE GO FROM HERE?

If you're reading this book, you are likely already part of the ecosystem: as a policymaker, funder, builder, educator, researcher, or advocate.

Here are five invitations for what you can do next:

1. Audit Your Role

Ask: How is my work shaping who gets credentialed, what gets credentialed, and who benefits? Am I centered in the work, or are the people most affected? Who is not in the room?

2. Name the Harms

Include space for harm tracking, grievances, and public documentation in all credentialing initiatives. Create harm registries. Make them accessible. Act on what you learn.

3. Share Power

Codesign with community partners from the start, not just for feedback at the end. Share decision-making authority. Share budgets. Share credit. Share failure.

4. Fund What's Missing

Support infrastructure for independent evaluation, multi-lingual access, community governance, and opt-out mechanisms. Fund the unsexy work: maintenance, translation, accessibility, dispute resolution.

5. Choose People Over Pilots

Design for depth, not just breadth. Build systems that work for one community well before scaling. Prioritize dignity over efficiency. Move at the speed of trust.

CLOSING QUESTIONS

We'll end with the questions we hope this book has helped you ask:

- ▶ Who benefits, truly, materially benefits, from this system?
- ▶ Who is credentialing the credentialers? Who audits the auditors?
- ▶ Who is watching the wallet builders?
- ▶ Who gets to refuse?
- ▶ What counts as harm?
- ▶ And who decides what counts?

CONCLUSION

If we want to build a world where people are seen, trusted, and free, we must start by building systems that reflect that vision.

This work is not neutral. Yet, it can be just.

Let's build with humility, with courage, and with the communities who've been teaching us all along.

APPENDICES

APPENDIX A: GLOSSARY OF TERMS

Award/awarder. The award is the actual achievement, qualification, or recognition being documented (the degree, certificate, badge, etc.). The awarder is the authoritative body that grants or confers that achievement. An award has the institutional authority to say someone has met specific requirements. Examples: a university's Board of Trustees (awarder) confers a Bachelor's degree (award); a state licensing board (awarder) grants a nursing license (award); an industry consortium (awarder) certifies a professional credential (award)

Badge. A digital symbol or signifier that represents the achievement of a specific skill, behavior, or learning outcome. Often issued through platforms like Credly or Badgr.

Blockchain. A decentralized, cryptographic system used to record data across many computers. In credentialing, it is used for secure verification and timestamping.

Credential. A learning, work, and/or identity asset that gives credential or credibility to who someone is, their experiences, or what someone can do. A credential attests to someone's qualifications, competencies, or identity. Includes degrees, certifications, licenses, and digital badges.

Credential Harms Registry. A proposed public resource to document grievances and harms experienced by individuals using or impacted by digital credential systems.

Data Self-Sovereignty. The right of individuals or communities to own and control how their personal data is collected, stored, used, and shared, especially across national or institutional boundaries.

Decentralized Identifier (DID). A globally unique identifier that is created and managed independently of a central registry, often used in self-sovereign identity systems.

Digital Wallet or Storage Container. A software application that stores and manages digital credentials, such as verifiable credentials or identity tokens, which can be shared with verifiers.

Evaluation-Industrial Complex. A critique of ecosystems where the same actors designing, implementing, and funding technologies are also responsible for evaluating them, leading to biased or incomplete assessments.

Human-Centered Design. A design approach that prioritizes the lived experience, needs, and agency of people, especially marginalized or excluded communities.

Interoperability. The ability of different systems, platforms, or tools to work together seamlessly, critical in the exchange and verification of digital credentials.

Issuer. The issuer is the organization or entity that creates and digitally signs a credential, asserting that a claim about a person is true. The issuer is cryptographically responsible for the credential's validity. Examples: a university issuing a degree; an employer issuing a certificate of employment; a training provider issuing a completion badge; a professional association issuing a license. Technical role: In systems like Verifiable Digital Credentials (VDCs), the issuer uses cryptographic keys to sign the credential, making it tamper-evident and verifiable.

Learning and Worker Record (LWR) A learning and/or work 'record' is a 'documented account of an individual's activities, achievements, skills, or performance that serves as evidence of their capabilities, experiences, or progress.' It includes their credentials as well as other information.

Open Badge Standard. An open standard for digital badges developed by Mozilla and maintained by IMS Global (1EdTech), allowing metadata to accompany digital credentials.

Participatory Governance. A system of decision-making where affected communities are actively involved in defining rules, oversight, and accountability mechanisms.

Revocation. The process of invalidating a credential after it has been issued, typically by the original issuer. Can be transparent or opaque, and often lacks holder appeal rights.

Self-Sovereign Identity (SSI). A model of digital identity that allows individuals to control their identity data and credentials, without reliance on centralized authorities.

Selective Disclosure. A privacy-enhancing feature that allows individuals to share only specific parts of a credential (e.g., date of birth yet not name).

Skills-Based Hiring. An employment practice that prioritizes a candidate's skills over formal credentials like degrees. Often facilitated by digital credentials and LWRs.

Tokenization (of Experience). Reducing complex human behaviors or achievements into simple digital markers (e.g., 'resilience' badges), often stripping away important context.

Usability. The ease and usefulness within which a person can interact with a system or tool. In credentialing, poor usability can lead to abandonment or harm.

Verifier. The entity (e.g., employer, institution) that checks the validity of a credential when it is presented by the holder.

Verifiable digital credentials (VDCs) are a data-rich, socio-technical systems of assets and infrastructures that embody particular values, power, and assumptions for record checking, confirmability, and traceability.

Wallet Builder. A company or organization that creates digital wallet tools to manage credentials. Often holds significant control over access, consent flows, and data architecture.

APPENDIX B: HISTORICAL TIMELINE OF LEARNING AND WORK RECORD KEEPING

Historical Timeline of Learning and Work Record Keeping

Period	Civilization	How	Why	Description	Ref
3500-3200 BCE	Ancient Mesopotamia (Sumerians)	Invented cuneiform script using wedge shaped marks on clay tablets with reed stylus.	Needed to document trade transactions and manage agricultural surplus.	Created foundation for all future writing systems and record keeping. Complex system with over 700 symbols that changed between cities and over time.	[1]
3100-3130 BCE	Ancient Egypt	Elite scribes trained in hieroglyphics, serving temples and pharaonic administration.	Maintain a centralized economic system and preserve religious and cultural values.	Helped Egyptian civilization maintain a huge centralized economic system. The literacy rate was only 1% so it was an extremely exclusive system.	[2]
2900-2334 BCE	Mesopotamian Scribal Schools (Edubba)	Students aged 8-22 learned cuneiform, mathematics, and administration in the *"House of Tablets."*	Train scribes for temple and palace administrative needs.	Ensured literature, science, and law were transmitted across generations. Only children of the upper class could afford tuition; mostly male students.	[3]

206 BCE-220 CE	*Han Dynasty China*	First emperor Liu Bang ordered candidates to register with character, appearance, and age recorded.	Create systematic government employment and promotion systems.	Established merit-based civil service. Limited to government officials only.	[4]
618-907 CE	*Tang Dynasty China*	Detailed employee files recording personal information, work. experience, job performance, and references_	Maintain efficient government bureaucracy and prevent corruption.	Created a comprehensive personnel management system still used today. Files could control major life decisions like marriage and travel.	[4]
11th-16th Century	*Medieval European Guilds*	Apprentice (7 years) → Journeyman → Master system with written agreements and masterpieces.	Protect industry from competition, maintain quality standards, increase political influence.	Mutual aid, quality control, professional mobility, education support. Became hereditary and exclusionary; set artificially high standards.	[5
1563	*English Statute of Artificers*	First national apprenticeship system: 7-year terms, maximum 3 apprentices per master.	Standardize training and prevent exploitation.	Created a legal framework for professional training. Only applied to trades that existed when law was passed.	[6]
1710-1811	**English Apprenticeship Records**	Official records kept when stamp duty was payable on indentures of apprenticeship.	Government revenue and legal protection.	Created systematic documentation of professional training. Many informal apprenticeships avoided the system to avoid taxes.	[7]

| 19th-20th Century | *Industrial Transition* | Guild power faded due to industrialization and rise of nation-states. | Mass production required standardized methods rather than guild secrets. | Enabled industrial scale production and innovation.

Many former handicraft workers were forced into manufacturing with less job security. | [8] |

© LWYL Studio.

Sources:

1. Edsitement; Cuneiform tablets reveal secrets of Mesopotamian payroll - ADP ReThink Q; The Life of a Scribe in Ancient Mesopotamia; Features - The World's Oldest Writing - Archaeology Magazine - May/June 2016

2. BrewminateBrewminate; Education in Ancient Civilizations; From scribes to software: A brief history of bookkeeping; A History of Education from the Ancient World to Today

3. Brewminate; Frayne - Scribal Education in Ancient Babylonia; The Life of a Scribe in Ancient Mesopotamia; From the Edubba: Education in Ancient Mesopotamia

4. Why Does China Still Keep a Secret File on Every Worker? | The World of Chinese

5. Guild | Trade Associations & Their Role in Medieval Europe | Britannica; Medieval Guilds - World History Encyclopedia; Encyclopedia BritannicaStudy.com

6. A short history of apprenticeships in England: from medieval craft guilds to the twenty-first century

7. Apprentices and masters - The National Archives

8. Guild - Wikipedia

APPENDIX C: TYPOLOGY OF LEARNING AND WORK RECORDS

Typology of Learning and Work Records

Type	Historical Version	Purpose	Content	Access and/or Control	Modern Version	Ref.
Student Progress Records	Mesopotamian edubba tablets showing cuneiform writing exercises and curriculum stages.	Track learning progression and skill development.	Individual student work, corrections, advancement through curriculum levels	Teachers and scribal schools	Transcripts, grade reports, learning portfolios	[1]
Professional Certification	Medieval guild masterpieces and journeyman certificates proving craft competency	Validate professional qualifications and skill mastery	Completed masterwork, peer approval, guild membership status	Guild masters and officials	Professional licenses, certifications, diplomas	[2]
Employment History	Tang Dynasty personnel files recording work experience, job performance assessments, and references	Document career progression and work performance	Previous positions, supervisor evaluations, career achievements	Government administrators	Résumés, employment records, performance reviews	[3]
Apprenticeship Agreements	English apprenticeship indentures documenting terms, duration, and completion	Formalize training relationships and obligations	Training period, master's responsibilities, completion requirements	Legal authorities and guild officials	Internship agreements, training contracts	[4]
Payroll and Wage Records	Mesopotamian clay tablets documenting worker wages, obligations, and rights between employers and workers	Track compensation and work arrangements	Payment rates, work assignments, hours, benefits	Temple and palace administrators	Payroll records, employment contracts	[5]

Educational Enrollment	Edubba school admission records for children of upper class and nobility	Control access to educational opportunities	Student identity, family background, tuition payment	Educational institutions	School enrollment records, admissions files	[6]
Skill Assessment	Sumerian curriculum records tracking progression through Tetrad and Decad compositions.	Measure competency and readiness for advancement	Specific skills mastered, knowledge areas completed	Educational authorities	Competency assessments, skill certifications	[7]
Guild Membership	Medieval guild rolls documenting member status, dues, and privileges	Control professional practice and maintain standards	Membership level, financial obligations, voting rights	Guild leadership	Professional association membership	[8]
Disciplinary Records	Sumerian "Schooldays" describing student punishments and behavioral issues	Document rule violations and corrective actions	Infractions, punishments administered, behavioral patterns	School administrators	Disciplinary files, conduct records	[9]
Teaching Credentials	Mesopotamian scribal qualifications for temple and palace instruction roles.	Authorize individuals to provide training	Subject expertise, teaching experience, institutional approval	Religious and governmental authorities	Teaching licenses, faculty credentials	[10]
Trade Secrets and/or Knowledge	Medieval guild records of protected techniques and methods.	Preserve and control specialized knowledge	Technical processes, trade secrets, quality standards	Guild masters only	Proprietary training materials, trade documentation	[11]
Social and/or Political Background	Chinese dang'an files including political affiliations and family background.	Assess trustworthiness and social standing	Political party membership, family history, social connections	State security apparatus	Background checks, security clearances	[12]

© LWYL Studio

Sources:

1. Eduba - Wikipedia

2. Guild | Trade Associations & Their Role in Medieval Europe | Britannica

3. Why Does China Still Keep a Secret File on Every Worker? | The World of Chinese

4. The National ArchivesHouse of Commons Library

5. Cuneiform tablets reveal secrets of Mesopotamian payroll - ADP ReThink Q

6. From the Edubba: Education in Ancient Mesopotamia

7. Frayne - Scribal Education in Ancient Babylonia

8. Medieval Guilds - World History Encyclopedia

9. School Days: Ancient Sumerian Satire and the Scribal Life Brewminate: A Bold Blend of News and Ideas

10. The Life of a Scribe in Ancient Mesopotamia

11. Medieval Guilds | Types, Hierarchy & Function - Lesson | Study.com

12. Why Does China Still Keep a Secret File on Every Worker? | The World of Chinese

APPENDIX D: VENDOR AND PRODUCT OVERVIEW

Vendor	Product Name	Powered By	Status	Focus
Accredible	Accredible	Accredible	Private Company	Certificates, badges, digital wallets
Arizona State University	Pocket Wallet	Digital Credential Consortium	University-led	Digital credentials and wallets
BCdiploma	BCdiploma	BCdiploma	Private Company	Blockchain-secured diplomas and academic credentials
CertifyMe	CertifyMe	Tech99 Innovations Pvt Ltd	Private Company	Digital credentials for education and corporate training
Pearson	Credly Acclaim	Pearson	Private Company	Digital credentials, badges, workforce credentials
Digital Bazaar	Veres Wallet	Digital Bazaar	Private Company	Digital wallets and verifiable credentials
Truvera (formerly Dock)	Truvera Dock Certs	Truvera	Private Company	Digital credentials
EBSCO	EBSCOed Digital Credential Wallet	EBSCO	Private Company	Library and education technology integration
Gobekli	Universal Talent Passports	Gobekli	Private Company	Talent passports and skills verification
Gradintelligence	GradIntel	Tribal Group (UK)	Part of Tribal Group	Digital Higher Education credentials and student records
Greenlight Credentials	Greenlight Credentials (Digital Locker & Credentials Wallet)	GreenLight Credentials LLC	Private Company	Blockchain-secured digital locker, lifelong learning records

iDatafy	SmartResume	iDatafy	Private Company	Verified resumes and employ-ment-linked credentials
Instructure	Parchment + Digitary	Parchment	Private Company	Academic credentialing, secure dig-ital records, Transcripts, diplomas, certifi-cates, credentials
Instructure	Canvas Credentials (formerly Badgr / Concentric Sky)	Instructure	Private Company	Open Badges and LMS-integrated cre-dentials (former-ly Badgr)
IQ4	IQ4 Wallet	IQ4	Private Company	Achievement wallets, skills passports
Learning Economy Foundation	Learn Card	Learning Economy Foundation	Nonprofit, hybrid structure in partnership with de-velopment house WeLibrary, LLC	Credentialing infrastructure, LearnCard wallet
Level Data	RANDA	RANDA Solutions	Private Company	Comprehensive learner records and credentials
Proof of Knowledge (POK)	POK	Optimism, Ethereum and Polygon	Private Company	Blockchain-based credentials
SchooLinks	SchooLinks Platform	SchooLinks	Private Company	K-12 college and career readiness, credential wal-lets, work-based learning
SpruceID	Credible, Verifier, SpruceKit	SpruceID	Private Company	Digital identi-ty credentials, mobile driver's li-censes, verifiable credentials
Territorium	TerritoriumCLR	Territorium	Private Company	Learning and employment records (LERs), skills transcripts
The SOLO Network	SOLO	Spark +	Private Company	Digital cre-dentials and interoperability

TrueCred	TrueCred	Accreditrust Technologies	Private Company	Digital badges, micro-credentials for workforce and education
Velocity Network Foundation	Velocity Career Wallet	Velocity blockchain	Nonprofit Consortium	Career identity and wallet infrastructure

Source: LWYL Studio (forthcoming, 2026)

APPENDIX E: EQUITY-INFORMED DESIGN FRAMEWORKS

The following frameworks and toolkits can support practitioners, researchers, and developers in designing digital credentialing systems that center equity, justice, and consent.

These approaches offer structured methods for community engagement, harm reduction, and ethical governance, particularly for historically marginalized populations. They also offer blueprints not just for inclusive design, yet for systems that are structurally accountable, liberatory in intention, and co-governed by those most impacted.

Integrating the frameworks into verifiable digital credentialing infrastructure isn't optional, it's ethical due diligence.

Name	Description	Concepts Principles
Data for Black Lives (D4BL) Justice-Informed Data Practices	A movement-building framework that advocates for data practices rooted in racial justice, self-determination, and the redistribution of power. Emphasizes community control, data stewardship, and the abolition of harmful surveillance systems.	▸ Data as protest and protection ▸ Abolitionist tech practices ▸ Reclaiming data for collective liberation

<u>Consentful Tech Project: HackBlossom</u>	A framework for building technologies that respect agency, boundaries, and affirmative consent, drawing from feminist theory, trauma-informed design, and digital security practices.	▶ Freely given, reversible, informed, enthusiastic, and specific (FRIES) consent ▶ Respectful defaults, visibility into data sharing ▶ Human empowerment and refusal-friendly design
<u>Design Justice Network: Principles and Toolkit</u>	A community-led network that centers marginalized voices in technology and design processes. Offers ten guiding principles for participatory design and collective accountability.	▶ We center the voices of those directly impacted ▶ We redistribute decision-making power ▶ We build community capacity and autonomy
Codesign Justice: Participatory Research and Community Stewardship	A practice rooted in decolonial and anti-oppressive methodologies that includes community stakeholders as full partners throughout the design and governance lifecycle.	▶ Community-led design sprints ▶ Advisory boards with power-sharing ▶ Long-term co-ownership models

<u>Our Data Bodies: Digital Defense Playbook</u>	A popular education tool for people navigating data collection, surveillance, and algorithmic harm in their daily lives. Developed by the Our Data Bodies project (Allied Media Projects, 2019).	▶ Workshop templates for community awareness ▶ Case studies of data harms ▶ Storytelling as analysis
<u>MyData Global: Human-Centric Data Governance</u>	An international initiative advocating for personal data control and interoperability, with an emphasis on transparency, trust frameworks, and ethical data exchange.	▶ Human-centricity ▶ Interoperability without surveillance ▶ Individual agency in data flows
<u>Equity-Centered Design Framework (EquityXDesign)</u>	Developed by equityXdesign and the National Equity Project, this approach blends human-centered design with critical consciousness.	▶ Acknowledge power and identity ▶ Design at the margins ▶ Interrupt inequitable patterns

APPENDIX F: CHECKLIST FOR PEOPLE-CENTERED AND ETHICAL DEPLOYMENT

This checklist for people-centered and ethical deployment is intended to guide both project teams and policymakers involved in the design, development, and/or deployment of Learning and Work Record (LWR) systems and their issuance as verifiable digital credentials (VDCs). It centers consent, equity, usability, and governance. It is recommended that it be used throughout the lifecyle of a project.

LWR/VDC People-Centered and Ethical Deployment Checklist

Consent and Control	√
Are holders informed of what the LWR/VDC contains, how it will be issued, and/or used?	
Can holders opt out of receiving, sharing, or storing their LWR/VDC without penalty?	
Is selective disclosure supported and available to holders so they may select the data they wish to share and with whom?	
Can the holder delete credentials from their wallet, and have them deleted from both the wallet and any server instances (.e.g., in cloud storage or archives)?	
Is holder consent engabled for third-party sharing and verification?	

Governance and Oversight	√
Is there a clearly defined governance model and oversight structure?	
Are all ecosystem groups represented in the governance structure and poilcy decisions, including learners and workers?	
Are issuer and verifier roles transparent and accountable?	

Is there an independent evaluation or audit process for all systems, protocls, and tools being tested, demonstrated, and/or deployed?	
Are there dispute resolution and harm redress mechanisms?	
Are there detailed procurement and evaluation protocols?	

Equity and Inclusion	✓
Was the system designed, developed, and deployed with involved and directly impacted communities?	
Does the LWR/VDC framework recognize non-institutional or formal learning and work (e.g., care work, volunteer, self-direct study)?	
Are usability and literacy needs, including multilingual access prioritized?	
Are exclusion risks assessed for people without regular access to phones, emails, or the internet?	
Is disaggregated impact data (e.g., race, disability, income) collected and reported?	

Usability and Experience	✓
Has the system been tested with learners and workers before launch?	
Are interfaces accessible (WCAG compliant, mobile-friendly)?	
Is credential metadata understandable by both humans and machines?	
Are holders given guidance on how to use and share their credentials?	
Are revocation, correction, and expiration clearly communicated?	

Procurement and Vendor Transparency	✓
Are vendor roles and responsibilities disclosed publicly?	
Is the procurement process open, competitive, and bias-aware?	
Are vendors evaluated based on usability, equity, and consent, not just compliance?	
Are any product or platform incentives aligned with learner/worker control and public good?	

Data and Privacy	✓
Are data flows mapped and documented for all parties?	
Is personal data minimized and stored only when necessary?	
Are logs (e.g., access, verification) controlled and transparent to the holder?	
Is privacy impact assessment conducted and publicly posted?	
Are retention policies defined, especially after a person leaves a job or school?	

Monitoring and Evaluation	✓
Are success metrics aligned with human outcomes (e.g., employment access, holder trust)?	
Are harms, complaints, or non-use documented and analyzed?	
Is a harm reporting mechanism in place for holders and issuers?	
Are evaluation results publicly shared and disaggregated?	
Is there a plan for iterative redesign based on findings?	

AUTHOR DETAILS

Kelly Page, PhD

Dr. Kelly Page is a social design ethnographer and inclusive usability designer committed to developing inclusive, accessible, and consent-driven innovations, as well as creating truly caring and inclusive cultures, leaders, and experiences

With a PhD in the *Psychology of Web Knowledge*, she has over 25 years of experience working at the intersection of social innovation, usability, and digital transformation. Her work explores the adoption and use of learning and earning technologies, digital credentialing, Web3, and inclusive and usable consent-driven design. She is the founder and CEO of LWYL Studio and an affiliate faculty member at the University of Colorado Denver.

Her research has been published in leading peer-reviewed journals and featured in The New York Times, Fast Company, and The Wall Street Journal. She advises emerging tech leaders, guides boards and product teams in building more inclusive product and service innovations, and serves as a research and learning fellow.

Books by Dr. Page include *'When Credentials Cause Harm'* and *'By CoDesign.'* To learn more, visit: drkellypage.com or lwylstudio.com or follow on social media at @drkellypage and @LWYLStudio.

www.ingramcontent.com/pod-product-compliance
Lightning Source LLC
Chambersburg PA
CBHW061023220326
41597CB00019BB/3062